Lévy Processes in Finance

Lévy Processes in Finance

Pricing Financial Derivatives

Wim Schoutens

Katholieke Universiteit Leuven, Belgium

WILEY

Other Wiley Editorial Offices

John Wiley & Sons Inc., 111 River Street, Hoboken, NJ 07030, USA

Jossey-Bass, 989 Market Street, San Francisco, CA 94103-1741, USA

Wiley-VCH Verlag GmbH, Boschstr. 12, D-69469 Weinheim, Germany

John Wiley & Sons Australia Ltd, 33 Park Road, Milton, Queensland 4064, Australia

John Wiley & Sons (Asia) Pte Ltd, 2 Clementi Loop #02-01, Jin Xing Distripark, Singapore 129809

John Wiley & Sons Canada Ltd, 22 Worcester Road, Etobicoke, Ontario, Canada M9W 1L1

Wiley also publishes its books in a variety of electronic formats. Some content that appears in print may
not be available in electronic books.

British Library Cataloguing in Publication Data

A catalogue record for this book is available from the British Library

ISBN 0-470-85156-2

Typeset in 10/12pt Times by T&T Productions Ltd, London.
Printed and bound in Great Britain by Biddles Ltd, Guildford, Surrey.
This book is printed on acid-free paper responsibly manufactured from sustainable forestry
in which at least two trees are planted for each one used for paper production.

To Ethel, Jente and Maitzanne

Contents

Preface   xi

Acknowledgements   xv

1   Introduction   1
   1.1   Financial Assets   1
   1.2   Derivative Securities   3
          1.2.1   Options   3
          1.2.2   Prices of Options on the S&P 500 Index   5
   1.3   Modelling Assumptions   7
   1.4   Arbitrage   9

2   Financial Mathematics in Continuous Time   11
   2.1   Stochastic Processes and Filtrations   11
   2.2   Classes of Processes   13
          2.2.1   Markov Processes   13
          2.2.2   Martingales   14
          2.2.3   Finite- and Infinite-Variation Processes   14
   2.3   Characteristic Functions   15
   2.4   Stochastic Integrals and SDEs   16
   2.5   Financial Mathematics in Continuous Time   17
          2.5.1   Equivalent Martingale Measure   17
          2.5.2   Pricing Formulas for European Options   19
   2.6   Dividends   21

3   The Black–Scholes Model   23
   3.1   The Normal Distribution   23
   3.2   Brownian Motion   24
          3.2.1   Definition   25
          3.2.2   Properties   26
   3.3   Geometric Brownian Motion   27

3.4   The Black–Scholes Option Pricing Model   28
       3.4.1   The Black–Scholes Market Model   29
       3.4.2   Market Completeness   30
       3.4.3   The Risk-Neutral Setting   30
       3.4.4   The Pricing of Options under the Black–Scholes Model   30

4   Imperfections of the Black–Scholes Model   **33**

4.1   The Non-Gaussian Character   33
       4.1.1   Asymmetry and Excess Kurtosis   33
       4.1.2   Density Estimation   35
       4.1.3   Statistical Testing   36
4.2   Stochastic Volatility   38
4.3   Inconsistency with Market Option Prices   39

5   Lévy Processes and OU Processes   **43**

5.1   Lévy Processes   44
       5.1.1   Definition   44
       5.1.2   Properties   45
5.2   OU Processes   47
       5.2.1   Self-Decomposability   47
       5.2.2   OU Processes   48
5.3   Examples of Lévy Processes   50
       5.3.1   The Poisson Process   50
       5.3.2   The Compound Poisson Process   51
       5.3.3   The Gamma Process   52
       5.3.4   The Inverse Gaussian Process   53
       5.3.5   The Generalized Inverse Gaussian Process   54
       5.3.6   The Tempered Stable Process   56
       5.3.7   The Variance Gamma Process   57
       5.3.8   The Normal Inverse Gaussian Process   59
       5.3.9   The CGMY Process   60
       5.3.10   The Meixner Process   62
       5.3.11   The Generalized Hyperbolic Process   65
5.4   Adding an Additional Drift Term   67
5.5   Examples of OU Processes   67
       5.5.1   The Gamma–OU Process   68
       5.5.2   The IG–OU Process   69
       5.5.3   Other Examples   70

6   Stock Price Models Driven by Lévy Processes   **73**

6.1   Statistical Testing   73
       6.1.1   Parameter Estimation   73
       6.1.2   Statistical Testing   74

6.2   The Lévy Market Model   76
       6.2.1   Market Incompleteness   77
       6.2.2   The Equivalent Martingale Measure   77
       6.2.3   Pricing Formulas for European Options   80
6.3   Calibration of Market Option Prices   82

7 Lévy Models with Stochastic Volatility   85
7.1   The BNS Model   85
       7.1.1   The BNS Model with Gamma SV   87
       7.1.2   The BNS Model with IG SV   88
7.2   The Stochastic Time Change   88
       7.2.1   The Integrated CIR Time Change   89
       7.2.2   The IntOU Time Change   90
7.3   The Lévy SV Market Model   91
7.4   Calibration of Market Option Prices   97
       7.4.1   Calibration of the BNS Models   97
       7.4.2   Calibration of the Lévy SV Models   98
7.5   Conclusion   98

8 Simulation Techniques   101
8.1   Simulation of Basic Processes   101
       8.1.1   Simulation of Standard Brownian Motion   101
       8.1.2   Simulation of a Poisson Process   102
8.2   Simulation of a Lévy Process   102
       8.2.1   The Compound Poisson Approximation   103
       8.2.2   On the Choice of the Poisson Processes   105
8.3   Simulation of an OU Process   107
8.4   Simulation of Particular Processes   108
       8.4.1   The Gamma Process   108
       8.4.2   The VG Process   109
       8.4.3   The TS Process   111
       8.4.4   The IG Process   111
       8.4.5   The NIG Process   113
       8.4.6   The Gamma–OU Process   114
       8.4.7   The IG–OU Process   115
       8.4.8   The CIR Process   117
       8.4.9   BNS Model   117

9 Exotic Option Pricing   119
9.1   Barrier and Lookback Options   119
       9.1.1   Introduction   119
       9.1.2   Black–Scholes Barrier and Lookback Option Prices   121
       9.1.3   Lookback and Barrier Options in a Lévy Market   123

9.2   Other Exotic Options   125
        9.2.1   The Perpetual American Call and Put Option   125
        9.2.2   The Perpetual Russian Option   126
        9.2.3   Touch-and-Out Options   126
9.3   Exotic Option Pricing by Monte Carlo Simulation   127
        9.3.1   Introduction   127
        9.3.2   Monte Carlo Pricing   127
        9.3.3   Variance Reduction by Control Variates   129
        9.3.4   Numerical Results   132
        9.3.5   Conclusion   134

10   Interest-Rate Models   135
10.1   General Interest-Rate Theory   135
10.2   The Gaussian HJM Model   138
10.3   The Lévy HJM Model   141
10.4   Bond Option Pricing   142
10.5   Multi-Factor Models   144

Appendix A   Special Functions   147
A.1   Bessel Functions   147
A.2   Modified Bessel Functions   148
A.3   The Generalized Hypergeometric Series   149
A.4   Orthogonal Polynomials   149
        A.4.1   Hermite polynomials with parameter   149
        A.4.2   Meixner–Pollaczek Polynomials   150

Appendix B   Lévy Processes   151
B.1   Characteristic Functions   151
        B.1.1   Distributions on the Nonnegative Integers   151
        B.1.2   Distributions on the Positive Half-Line   151
        B.1.3   Distributions on the Real Line   152
B.2   Lévy Triplets   153
        B.2.1 γ 153
        B.2.2   The Lévy Measure $\nu(\mathrm{d}x)$ 154

Appendix C   S&P 500 Call Option Prices   155

References   157

Index   165

Preface

The story of modelling financial markets with stochastic processes began in 1900 with the study of Bachelier (1900). He modelled stocks as a Brownian motion with drift. However, the model had many imperfections, including, for example, negative stock prices. It was 65 years before another, more appropriate, model was suggested by Samuelson (1965): geometric Brownian motion. Eight years later Black and Scholes (1973) and Merton (1973) demonstrated how to price European options based on the geometric Brownian model. This stock-price model is now called the Black–Scholes model, for which Scholes and Merton received the Nobel Prize for Economics in 1997 (Black had already died).

It has become clear, however, that this option-pricing model is inconsistent with options data. Implied volatility models can do better, but, fundamentally, these consist of the wrong building blocks. To improve on the performance of the Black–Scholes model, Lévy models were proposed in the late 1980s and early 1990s, since when they have been refined to take account of different stylized features of the markets.

This book is concerned with the pricing of derivative securities in market models based on Lévy processes. Financial mathematics has recently enjoyed considerable prestige as a result of its impact on the finance industry. The theory of Lévy processes has also seen exciting developments in recent years. The fusion of these two fields of mathematics has provided new applied modelling perspectives within the context of finance and further stimulus for the study of problems within the context of Lévy processes.

This book is aimed at people working in the areas of mathematical finance and Lévy processes, with the intention of convincing the former that the rich theory of Lévy processes can lead to tractable and attractive models that perform significantly better than the standard Black–Scholes model. For those working with Lévy processes, we hope to show how the objects they study can be readily applied in practice.

We have taken great care not to use too much esoteric mathematics, nor to get too involved in technicalities, nor to give involved proofs. We focus on the ideas, and the intuition behind the modelling process and its applications. Nevertheless, the processes involved in the modelling are described very accurately and in great detail. These processes lie at the heart of the theory and it is very important to have a clear view of their properties.

This book is organized as follows. In Chapter 1 we introduce the phenomena we want to model: financial assets. Then we look at some of the basic modelling assumptions that are used throughout the book. Special attention is paid to the no-arbitrage assumption.

In Chapter 2, we recall the basics of mathematical finance in continuous time. We briefly discuss stochastic processes in continuous time together with stochastic integration theory. The main focus is on different pricing methods, arbitrage-free and (in)complete markets.

Chapter 3 introduces the famous Black–Scholes model. We give an overview of the model together with its basic properties and then we have a close look at the pricing formulas under this model.

Chapter 4 discusses why the Black–Scholes model is not such an appropriate model. First, we argue that the underlying Normal distribution is not suitable for the accurate modelling of stock-price behaviour. Next, we show that the model lacks the important feature of stochastic volatility. These imperfections are shown based on historical data. Moreover, we show that the model prices do not correspond as they should to market prices. The above discussed imperfections cause this discrepancy between model and market prices.

Chapter 5 is devoted to the main ingredients of the more sophisticated models introduced later on. We give an overview of the theory of Lévy processes and the theory of Ornstein–Uhlenbeck processes (OU processes). Lévy processes are based on infinitely divisible distributions. A subclass of these distributions is self-decomposable and leads to OU processes. A whole group of very popular examples of these processes is looked at in detail. Besides stating the defining equations, we also look at their properties.

In Chapter 6, for the first time we use non-Brownian Lévy processes to describe the behaviour of a stock-price process. We discuss the Lévy market model, in which the stock price follows the exponential of a Lévy process. The market models proposed are no longer complete and an equivalent martingale measure has to be chosen. Comparing model prices with market prices demonstrates that Lévy models are a significant improvement on the Black–Scholes model, but that they still fall short.

It is in Chapter 7 that we introduce stochastic volatility in our models. This can be done in several ways. We could start from the Black–Scholes model and make the volatility parameter involved itself stochastic. We focus on models where this volatility follows an OU process. Or we could introduce stochastic volatility by making time stochastic. If time goes fast, the market is nervous. If time goes slow, volatility is low. This technique can be applied not only in the Black–Scholes model but also in the Lévy market model. Different choices of stochastic time are considered: the rate of change of time can be described by the classical mean-reverting Cox–Ingersoll–Ross (CIR) stochastic process, but the OU processes are also excellent candidates.

Chapter 8 discusses simulation techniques. Attention is paid to the simulation of Lévy processes. Here we use the general method of approximating a Lévy process by compound Poisson processes. Particular examples of Lévy processes can be simulated by making use of some of their properties. When the process is a time change or a subordination of a simpler process, this method can be of special advantage. Paths from OU processes can be simulated by using a series representation or by the classical Euler scheme approximation.

Chapter 9 gives an overview of the pricing of exotic options under the different models. Typically, more or less explicit solutions are available under the Black–Scholes model. The situation worsens, however, under the Lévy market model. For barrier and lookback options, some results are available; however, the explicit calculations of prices in these cases are highly complex. Multiple integrals and inversion techniques are needed for numerical evaluation. In the even more advanced stochastic volatility models, prices can only be estimated by Monte Carlo simulations. In contrast with the Black–Scholes model, no closed formulas are available for the pricing of exotic options such as barrier and lookback options. The simulation techniques from Chapter 8 are used intensively here.

Finally, Chapter 10 focuses on interest-rate modelling. We have so far assumed that interest rates are constant; in practice, of course, they are not. We follow the Heath–Jarrow–Merton approach and model the entire yield curve. As with the stock-price models, the underlying Brownian motion does not describe the empirical behaviour as it should. In order to make the model more realistic, we replace it again by a more flexible Lévy process.

Acknowledgements

I owe special thanks to my scientific mentor, J. L. Teugels, for his continuous interest, help and encouragement. I am also extremely grateful to everyone who made this book possible. I thank Philippe Jacobs at KBC Bank for useful discussions, and for carefully reading and correcting preliminary versions of the typescript and for the helpful suggestions. I thank the members of Wim Allegaert's group at KBC Asset Management for useful discussions, and for their support and interest. Erwin Simons and Jurgen Tistaert at ING Financial Markets helped a lot through their useful discussions and their numerous and detailed suggestions for the correction of errors and other improvements.

I also thank the Fund for Scientific Research – Flanders (Belgium), the Katholieke Universiteit Leuven and the EURANDOM Institute for creating the environment in which I could freely do research. The latter is especially acknowledged for giving me the opportunity to co-organize the workshop on the Applications of Lévy processes in Financial Mathematics. This workshop was the basis of many interesting contacts.

I thank David Nualart and the organizers of several MaPhySto events (Ken-Iti Sato, Thomas Mikosch, Elisa Nicolato, Goran Peskir and Ole E. Barndorff-Nielsen) for their hospitality.

I also thank the organizers of the Master of Statistics Programme and the Master in Actuarial Science Programme at the Katholieke Universiteit Leuven for giving me the opportunity to teach courses on mathematical finance. Part of this book is based on those lecture notes. And I thank the students attending the lectures over the years for their illuminating suggestions.

Furthermore, I thank Dilip Madan, Andreas Kyprianou, Giovanni Vanroelen, Stijn Symens, Nick Bingham and Marc Yor for helping me in different ways. I also thank the staff at Wiley and at T&T Productions for making all this possible and for their warm collaboration.

Last and most, I want to thank my wife, Ethel, my son, Jente, and my daughter, Maitzanne, for their infinite love.

Wim Schoutens
Leuven

1

Introduction

Before we begin the detailed description of the models and their ingredients, we first focus on the financial markets we want to model together with the main group of underlying assets and their derivatives we want to price. We often closely follow Bingham and Kiesel (1998).

1.1 Financial Assets

The main goal of this book is to find attractive, useful and tractable models for financial time series. The models we present in the main part of the book (Chapters 1–9) are suitable for describing the stochastic behaviour of mainly stocks and (stock) indices. Other models can be set up for other assets such as commodities, currencies and interest rates (see Chapter 10).

Stocks

The basis of modern economic life is the company owned by its shareholders; the shares provide partial ownership of the company, pro rata with investment. Shares are issued by companies to raise funds. They have value, reflecting both the value of the company's real assets and the earning power of the company's dividends. Stock is the generic term for assets held in the form of shares. With publicly quoted companies, shares are quoted and traded on a stock exchange or bourse. (Some say the term 'bourse' derives from the merchant family, Van der Burse.) Stock markets date back to at least 1531, when one was started in Antwerp, Belgium. Today there are over 150 stock exchanges throughout the world.

Indices

An index tracks the value of a basket of stocks (FTSE100, S&P 500, Dow Jones Industrial, NASDAQ Composite, BEL20, EUROSTOXX50, etc.), bonds, and so on. Derivative instruments on indices may be used for hedging (covering against risk) if no derivative instruments on a particular asset in question are available and if the

Lévy Processes in Finance W. Schoutens
© 2003 John Wiley & Sons, Ltd ISBN: 0-470-85156-2

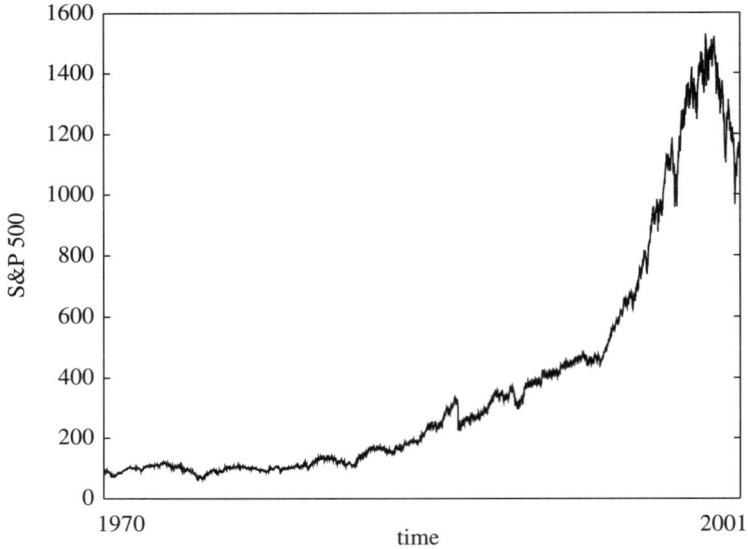

Figure 1.1   The S&P 500 Index from 1970 until the end of 2001.

correlation in movement between the index and the asset is significant. Furthermore, institutional funds (such as pension funds), which manage large, diversified stock portfolios, try to mimic particular stock indices and use derivatives on stock indices as a portfolio management tool. On the other hand, a speculator may wish to bet on a certain overall development in a market without exposing him/herself to a particular asset.

We will illustrate our theory with the S&P 500 Index, which is a weighted average of the main 500 American stocks. Figure 1.1 shows the daily prices of this index over a period of more than 30 years.

Dividends

Divivends are an individual share of earnings distributed among stockholders of a corporation or company in proportion to their holdings. Dividends are usually payable in cash, although sometimes distributions are made in the form of additional shares of stocks.

In our analysis, we first use nondividend-paying assets as the underlying assets. Later on in Section 2.6 we will indicate how to incorporate dividend payments into the models.

The Stock Price Process

We will model the price process of our asset (a stock or a index) by a continuous-time process. Throughout, we denote this asset price process (often referred to as the stock-price process) by $S = \{S_t, t \geq 0\}$; S_t gives us the price at time $t \geq 0$.

In order to allow for the comparison of investments in different securities, it is natural to look at the relative price changes (returns) over a time $s > 0$:

$$\frac{S_{t+s} - S_t}{S_t}.$$

For several reasons, most authors in the financial literature prefer working with logarithmic returns (or log returns) instead:

$$\log(S_{t+s}) - \log(S_t).$$

One reason is that the log returns over a period of length $k \times s$ are then the sum of the log returns of k periods of length s:

$$(\log(S_{t+s}) - \log(S_t)) + (\log(S_{t+2s}) - \log(S_{t+s})) + \cdots$$
$$+ (\log(S_{t+ks}) - \log(S_{t+(k-1)s})) = \log(S_{t+ks}) - \log(S_t).$$

Another reason is that in most models the stock price S_t will be modelled by an exponential of some basic stochastic process. Actually, for continuous-time processes, returns with continuous compounding log returns are the natural choice. Note that in what follows we also take continuously compounded interest rates in the model for the riskless asset (the bank account).

1.2 Derivative Securities

After finding an acceptable model for the price process of our asset, the next step is to price financial derivatives on the underlying asset (or simply the underlying).

Contingent Claims

Intuitively, a 'derivative security', or derivative for short, is a security whose value depends on the value of other more basic underlying securities. To be more precise, a derivative security, or *contingent claim*, is a financial contract whose value at expiration date T (more briefly, expiry) is determined exactly by the price process of the underlying financial assets (or instruments) up to time T.

Types of Derivatives

Derivative securities can be grouped under three general headings: options, forwards and futures and swaps. In this book, we will deal with options, although our pricing techniques may be readily applied to forwards, futures and swaps as well.

1.2.1 Options

An option is a financial instrument giving one the *right but not the obligation* to make a specified transaction at (or by) a specified date at a specified price.

Options are thus privileges sold by one party to another. The right is granted by the person who sells the option. The person who sells the option is called the seller or writer of the option. The person who buys the option is called the option buyer.

Option Types

Many different types of option exist. We give here the basic types. *Call* options give one the right to buy. *Put* options give one the right to sell.

European options give one the right to buy/sell on the specified date, the expiry date, when the option expires or matures.

The European call and put options, since they are so basic, are known as the *plain vanilla* options.

The more involved options are mostly called exotic. *American* options give one the right to buy/sell at any time prior to or at expiry. *Asian* options depend on the average price over a period. *Lookback* options depend on the maximum or minimum price over a period, and *barrier* options depend on some price level being attained.

Strike and Payoff Function

The price at which the transaction to buy/sell the underlying, on/by the expiry date (if exercised), is made is called the *exercise price* or the *strike price*. We usually denote the strike price by K, the initial time (when the contract between the buyer and the seller of the option is struck) by $t = 0$, and the expiry or final time by $t = T$.

The *payoff* of an option is its value at expiry. For a European call option with a strike price K, the payoff is

$$K = \begin{cases} S_T - K & \text{if } S_T > K, \\ 0 & \text{otherwise.} \end{cases}$$

We can also write this more concisely as $(S_T - K)^+$. If in this case $S_t > K$, the option is *in the money*; if $S_t = K$, the option is said to be *at the money*; and if $S_t < K$, the option is *out of the money*.

Advantages and Disadvantages of Using Options

Option strategies can be very risky. But, if used in the right way, options can be helpful: they allow you to drastically increase your leverage in a stock. This is a powerful feature for investors who follow speculative strategies. Options can also be a useful tool in hedging against unfavourable market movements. However, using options to speculate requires a close watch on open positions and a higher tolerance for risk than investing in stocks. Handling options correctly requires more than just a basic knowledge of the stock market. Options can be quite complicated and if you lack proper knowledge, you run the risk of losing a great deal of money.

The History of Contingent Claims

It is not known exactly when the first contingent claim contracts were used. We know that the Romans and Phoenicians used such contracts in shipping, and there is also evidence that Thales, a mathematician and philosopher in ancient Greece, used such contracts to secure a low price for olive presses in advance of the harvest. Thales had reason to believe the olive harvest would be particularly strong. During the off-season, when demand for olive presses was almost non-existent, he acquired rights – at a very low cost – to use the presses the following spring. Later, when the olive harvest was in full-swing, Thales exercised his option and proceeded to rent the equipment to others at a much higher price.

In Holland, trading in tulip derivatives blossomed during the early 1600s. At first, tulip dealers used contracts (call options) to make sure they could secure a reasonable price to meet demand. At the same time, tulip growers used other contracts (put options) to ensure an adequate selling price. However, it was not long before speculators joined the mix and traded these contracts for profit. Unfortunately, when the market crashed, many speculators failed to honour their agreements. The consequences for the economy were devastating.

Markets

Financial derivatives are basically traded in two ways: on organized exchanges and over-the-counter (OTC).

In 1973, the Chicago Board Options Exchange (CBOE) began trading in options on some stocks. The first listed options to be traded were call options; they were written on 16 different stocks on 26 April 1973. After repeated delays by the Securities and Exchange Commission (SEC), put trading finally began in 1977.

The CBOE's first home was actually a smoker's lounge at the Chicago Board of Trade. After achieving a first-day volume of 911 contracts, the average daily volume rocketed to over 20 000 the following year. Since then, the growth of options has continued its explosive trajectory.

Organized exchanges are subject to regulatory rules and require a certain degree of standardization of the traded instruments (strike price, maturity dates, size of contract, etc.).

OTC trading takes place between various commercial and investments banks, such as Goldman Sachs, Citibank, Morgan Stanley, Deutsche Bank, etc. Deals are made directly between the traders without a centralizing office or exchange. There are virtually no restrictions on the characteristics of the possible deals.

1.2.2   Prices of Options on the S&P 500 Index

Throughout this book we will test our models against a set of standard vanilla call options on the S&P 500 Index.

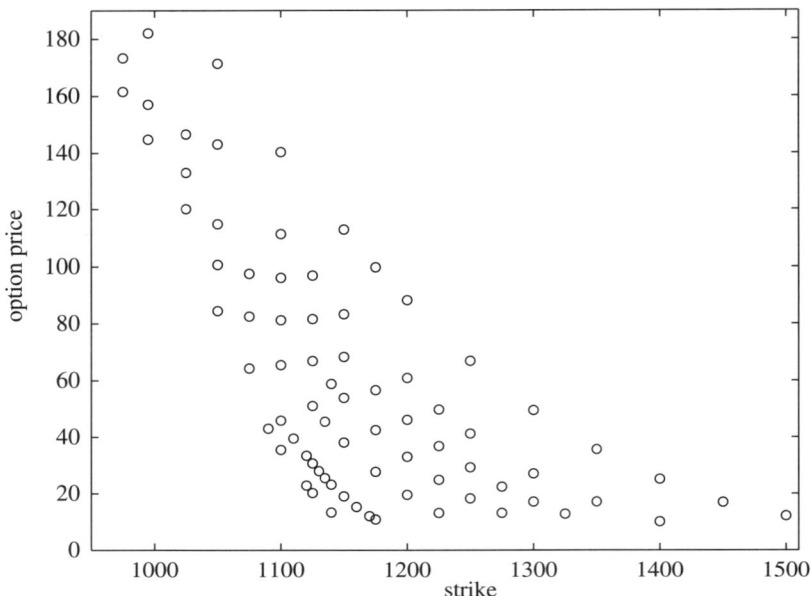

Figure 1.2 S&P 500 market option prices.

Option Prices Dataset

The dataset consists of 77 mid-prices of a set of European call options on the S&P 500 Index at the close of the market on 18 April 2002. On this day the S&P 500 closed at 1124.47. Since, by the put–call parity (see Section 1.4), the price of a put option can be calculated from the price of the call option with the same strike and maturity, and vice versa, we include in our set only call option prices. The exact prices can be found in Appendix C.

The option prices can be visualized as in Figure 1.2, which shows several series of call options. The upper series consists of options with the highest time to maturity corresponding to options expiring in December 2003. The inner series consists of options expiring in May 2002, June 2002, September 2002, December 2002, March 2003 and finally in June 2003.

We will calibrate different models to this set. The market prices are always denoted by a circle and later on the model prices will be denoted by a plus sign. It is the goal to calibrate the model such that our plus signs shoot right through the middle of the corresponding circles.

The parameters coming out of the calibration procedure resemble the current market view on the asset. Here we do not explicitly take into account any historical data. All the necessary information is contained in today's option prices, which we observe in the market. Using the available pricing techniques or by Monte Carlo simulations, this method is useful for pricing derivatives, such as OTC options, whose prices are not available in the market and for finding mispricings in a set of European vanilla options.

APE, AAE, RMSE and ARPE

For comparative purposes, we compute the average absolute error as a percentage of the mean price. This statistic, which we will denote by APE, is an overall measure of the quality of fit:

$$\text{APE} = \frac{1}{\text{mean option price}} \sum_{\text{options}} \frac{|\text{market price} - \text{model price}|}{\text{number of options}}.$$

Other measures which also give an estimate of the goodness of fit are the average absolute error (AAE), the average relative percentage error (ARPE) and the root-mean-square error (RMSE):

$$\text{AAE} = \sum_{\text{options}} \frac{|\text{market price} - \text{model price}|}{\text{number of options}},$$

$$\text{ARPE} = \frac{1}{\text{number of options}} \sum_{\text{options}} \frac{|\text{market price} - \text{model price}|}{\text{market price}},$$

$$\text{RMSE} = \sqrt{\sum_{\text{options}} \frac{(\text{market price} - \text{model price})^2}{\text{number of options}}}.$$

Typically, we estimate the model parameters by minimizing the root-mean-square error between the market and model prices.

1.3 Modelling Assumptions

Throughout we make use of a bank account. Our market model consists of this bank account (riskless) and one financial asset (risky), a stock or an index. We will discuss contingent claim pricing in an idealized case.

The Riskless Bank Account

First, we present a rule, which is valid throughout except for Chapter 10, for the bank account. The market dictates that there is a fixed interest rate $r \geqslant 0$. We can deposit money and borrow money on this same continuously compounded interest rate r. This means that 1 currency unit in the bank account at time 0 will give rise to $\exp(rt)$ currency units at time $t > 0$. Similarly, if we borrow 1 currency unit now, we will have to pay back $\exp(rt)$ currency units at time $t > 0$. Or, equivalently, if we now borrow $\exp(-rt)$ currency units, we will have to pay back 1 currency unit t time later.

Frictionless Market Assumptions

Second, we impose some rules for the risky asset. We will not allow market friction; there is no default risk, agents are rational and there is no arbitrage. In more concrete terms, this means

- no transaction costs (e.g. broker's commission),

- no bid/ask spread,

- perfect liquid markets,

- no taxes,

- no margin requirements,

- no restrictions on short sales,

- no transaction delays,

- market participants act as price takers,

- market participants prefer more to less.

The relaxation of all these assumptions is the subject of ongoing research.

We thus develop the theory of an ideal – frictionless – market in order to focus on the irreducible essentials of the theory and as a first-order approximation to reality.

The risk of failure of a company – bankruptcy – is inescapably present in its economic activity. The Enron bankruptcy and the WorldCom debacle make this very clear. Moreover, these risks also appear at the national level: quite apart from war, recent decades have seen default of interest payments of international debt, or the threat of it (for example, the 1998 Russian or the 2001 Argentinian crises). We ignore default risk for simplicity while developing our understanding of the principal aspects of the market.

We assume financial agents to be price takers, not price makers. The market is not influenced by the trading of individuals. This implies that even large amounts of trading in a security by one agent does not influence the security's price. Hence, agents can buy or sell as much of any security as they wish without changing the security's price. Moreover, we assume that brokers accept all trades. (In reality, they may want to work in round numbers.)

To assume that market participants prefer more to less is a very weak assumption on the preferences of market participants.

We consider a frictionless security market in which two assets are traded continuously. Investors are allowed to trade continuously up to some fixed finite planning horizon T, where all economic activity stops.

Finally, we mention the special character of the no-arbitrage assumption. It is the basis for the arbitrage pricing technique that we shall use, which we discuss in more detail below.

1.4 Arbitrage

In this section we follow Bingham and Kiesel (1998).

The essence of arbitrage is that with no initial capital it should not be possible to make a profit without exposure to risk. Were it possible to do so, arbitrageurs would do so, in unlimited quantity, using the market as a money-pump to extract arbitrarily large quantities of riskless profit. This would, for instance, make it impossible for the market to be in equilibrium.

The Put–Call Parity

Next, we will use arbitrage-based arguments to deduce a fundamental relation between put and call options, the so-called put–call parity. This relation is independent of the model that is assumed for the stock-price behaviour. It is a model-independent result based on the no-arbitrage assumption.

Suppose there is a nondividend-paying stock with value S_t at time t, with European call and put options on it, with values C_t and P_t, respectively, both with expiry time T and strike price K. Consider a portfolio consisting of one stock, one put and a short position in one call (the holder of the portfolio has written the call); write Π_t for the value at time t of this portfolio. So,

$$\Pi_t = S_t + P_t - C_t.$$

Recall that the payoffs at expiry are

$$C_T = \max\{S_T - K, 0\} = (S_T - K)^+, \qquad \text{for the call,}$$
$$P_T = \max\{K - S_T, 0\} = (K - S_T)^+, \qquad \text{for the put.}$$

Hence, for the above portfolio we get, at time T, the payoff

$$\Pi_T = \begin{cases} S_T + 0 - (S_t - K) = K & \text{if } S_T \geqslant K, \\ S_T + (K - S_t) - 0 = K & \text{if } S_T \leqslant K. \end{cases}$$

This portfolio thus guarantees a payoff K at time T. The riskless way to guarantee a payoff K at time T is to deposit $K \exp(-r(T - t))$ in the bank at time t and do nothing (we assume continuously compounded interest here). Under the assumption that the market is arbitrage-free, the value of the portfolio at time t must therefore be $K \exp(-r(T - t))$, since it acts as a synthetic bank account and any other price will offer arbitrage opportunities.

Indeed, if the portfolio is offered for sale at time t too cheaply, at price

$$\Pi_t < K \exp(-r(T - t)),$$

we can buy it, borrow $K \exp(-r(T - t))$ from the bank, and pocket a positive profit,

$$K \exp(-r(T - t)) - \Pi_t > 0.$$

Table 1.1   The case $\Pi_t < K \exp(-r(T - t))$.

Transactions	Current cash flow	Value at expiry $S_T < K$	Value at expiry $S_T \geqslant K$
buy 1 stock	$-S_t$	S_T	S_T
buy 1 put	$-P_t$	$K - S_T$	0
write 1 call	C_t	0	$-S_T + K$
borrow	$K \exp(-r(T - t))$	$-K$	$-K$
total	$K \exp(-r(T - t)) - S_t - P_t + C_t > 0$	0	0

At time T our portfolio yields K, while our bank debt has grown to K. We clear our cash account – using the one to pay off the other – thus locking in our earlier profit, which is riskless.

If on the other hand the portfolio is priced at time t at too a high price, at price $\Pi_t > K \exp(-r(T - t))$, we can do the exact opposite. We sell the portfolio short, that is, we buy its negative: buy one call, write one put, sell short one stock, for Π_t, and invest $K \exp(-r(T - t))$ in the bank account, pocketing a positive profit $\Pi_t - K \exp(-r(T - t)) > 0$. At time T, our bank deposit has grown to K, and again we clear our cash account – using this to meet our obligation K on the portfolio we sold short, again locking in our earlier riskless profit.

Arbitrage Table

We illustrate the above with a so-called arbitrage table (see Table 1.1). In such a table we simply enter the current value of a given portfolio and then compute its value in all possible states of the world when the portfolio is cashed in.

Thus the rational price for the portfolio at time t is exactly $K \exp(-r(T - t))$. Any other price presents arbitrageurs with an arbitrage opportunity.

Therefore, we have the following put–call parity between the prices of the underlying asset and its European call and put options with the same strike price and maturity on stocks that pay no dividends:

$$S_t + P_t - C_t = K \exp(-r(T - t)).$$

In Chapter 2, we will generalize this result for stocks which pay dividends.

2

Financial Mathematics in Continuous Time

This chapter summarizes the main results for continuous-time and continuous-variable processes in the context of finance. More detailed introductions can be found in, for example, Bingham and Kiesel (1998), Elliot and Kopp (1999), Hunt and Kennedy (2000), Shiryaev (1999) or Protter (2001).

It should be noted that, in practice, we do not observe stock prices following continuous-variable, continuous-time processes. Stock prices are restricted to discrete values (often multiples of 0.01 euros, dollars, etc.) and changes can be observed only when the exchange is open. Nevertheless, the continuous-variable, continuous-time process proves to be a useful model for many purposes.

2.1 Stochastic Processes and Filtrations

Probability Space and Filtrations

We assume a fixed finite planning horizon T. First, we need a probability space $(\Omega, \mathcal{F}, P)$. Ω is the set of all the possible outcomes that we are interested in. $\mathcal{F}$ is a sigma-algebra (a family of subsets of Ω closed under any countable collection of set operations) containing all sets for which we want to make a statement on; P gives the probability that an event in such a set of $\mathcal{F}$ will happen.

We call a probability space *P-complete* if for each $B \subset A \in \mathcal{F}$ such that $P(A) = 0$, we have that $B \in \mathcal{F}$. If we start with a probability space $(\Omega, \tilde{\mathcal{F}}, P)$, there exists a procedure to construct the so-called P-completion $(\Omega, \mathcal{F}, P)$, which is a complete probability space. We take $\mathcal{F} = \sigma(\tilde{\mathcal{F}} \cup \mathcal{N})$, where

$$\mathcal{N} = \{B \subset \Omega : B \subset A \text{ for some } A \in \tilde{\mathcal{F}}, \text{ with } P(A) = 0\}$$

and $\sigma(\mathcal{C})$ is the smallest sigma-algebra on Ω containing $\mathcal{C}$. We will always work with a complete probability space.

Lévy Processes in Finance   W. Schoutens
© 2003 John Wiley & Sons, Ltd   ISBN: 0-470-85156-2

Moreover, we equip our probability space $(\Omega, \mathcal{F}, P)$ with a *filtration*. A filtration is a nondecreasing family $\mathbb{F} = (\mathcal{F}_t, 0 \leqslant t \leqslant T)$ of sub-σ-algebras of $\mathcal{F}$:

$$\mathcal{F}_s \subset \mathcal{F}_t \subset \mathcal{F}_T \subset \mathcal{F} \quad \text{for } 0 \leqslant s < t \leqslant T;$$

here $\mathcal{F}_t$ represents the information available at time t, and the filtration $\mathbb{F} = (\mathcal{F}_t, 0 \leqslant t \leqslant T)$ represents the information flow evolving with time.

In general, we assume that the *filtered probability space* $(\Omega, \mathcal{F}, P, \mathbb{F})$ satisfies the following 'usual conditions'.

(a) $\mathcal{F}$ is P-complete.

(b) $\mathcal{F}_0$ contains all P-null sets of Ω. This means intuitively that we know which events are possible and which are not.

(c) $\mathbb{F}$ is right-continuous, i.e. $\mathcal{F}_t = \bigcap_{s>t} \mathcal{F}_s$; a technical condition.

If we start with a filtered probability space $(\Omega, \bar{\mathcal{F}}, P, \bar{\mathbb{F}})$, there exists a procedure to construct the so-called usual P-augmentation $(\Omega, \mathcal{F}, P, \mathbb{F})$, which satisfies the usual conditions. We take $\mathcal{F}$ to be equal to the P-completion of $\bar{\mathcal{F}}$ and set, for all $0 \leqslant t \leqslant T$,

$$\mathcal{F}_t = \bigcap_{s>t} \sigma(\bar{\mathcal{F}}_s \cup \mathcal{N}).$$

We will always work with filtered probability spaces which satisfy the usual conditions.

Stochastic Processes

A *stochastic process*, $X = \{X_t, 0 \leqslant t \leqslant T\}$, is a family of random variables defined on a complete probability space, $(\Omega, \mathcal{F}, P)$. We say that X is adapted to the filtration $\mathbb{F}$, or $\mathbb{F}$-*adapted*, if X_t is $\mathcal{F}_t$-measurable (we denote this by $X_t \in \mathcal{F}_t$) for each t: thus X_t is known at time t.

We say X is $\mathbb{F}$-*predictable* if $X_t \in \mathcal{F}_{t-} = \bigcup_{s<t} \mathcal{F}_s$ (i.e. X_t is $\mathcal{F}_{t-}$-measurable) for each t: thus X_t is known strictly before time t.

Starting with a stochastic process X on a complete probability space $(\Omega, \mathcal{F}, P)$, we call $\mathbb{F}^X = \{\mathcal{F}_t^X, 0 \leqslant t \leqslant T\}$ the natural filtration of X if it is the P-augmentation of the filtration $\bar{\mathbb{F}}^X = \{\bar{\mathcal{F}}_t, 0 \leqslant t \leqslant T\}$, where for each $0 \leqslant t \leqslant T$, $\bar{\mathcal{F}}_t^X$ is the smallest sigma-algebra such that X_t is $\bar{\mathcal{F}}_t^X$-measurable. It is thus the 'smallest' filtration (satisfying the usual conditions) containing all the information that can be observed if we watch X evolve through time.

Learning During the Flow of Time

The concept of filtration is not that easy to understand. We start by explaining the idea of filtration in a very idealized situation. We will consider a stochastic process X which starts at some value, say, zero. It will remain there until time $t = 1$, at which it can jump with positive probability to the value a or to a different value b. The process will stay at that value until time $t = 2$, at which it will jump again with positive probability to two different values: c and d, say, if the process was at state a at time $t = 1$, and f and g, say, if the process was at state b at time $t = 1$. From then on the process will stay at the same value. Ω consists of all possible paths the process can follow, i.e. all possible outcomes of the experiment. We will denote the path $0 \to a \to c$ by ω_1; similarly, the paths $0 \to a \to d, 0 \to b \to f$ and $0 \to b \to g$ are denoted by ω_2, ω_3 and ω_4, respectively. So, we have $\Omega = \{\omega_1, \omega_2, \omega_3, \omega_4\}$. We set here $\mathcal{F} = \mathcal{D}(\Omega)$, the set of all subsets of Ω.

In this situation, the natural filtration of X will be the following flow of information:

$$\begin{aligned}
\mathcal{F}_t &= \{\emptyset, \Omega\}, & 0 \leqslant t < 1; \\
\mathcal{F}_t &= \{\emptyset, \Omega, \{\omega_1, \omega_2\}, \{\omega_3, \omega_4\}\}, & 1 \leqslant t < 2; \\
\mathcal{F}_t &= \mathcal{D}(\Omega) = \mathcal{F}, & 2 \leqslant t \leqslant T.
\end{aligned}$$

To each of the filtrations given above, we associate, respectively, the following partitions (i.e. the finest possible one) of Ω:

$$\begin{aligned}
\mathcal{P}_0 &= \{\Omega\}, & 0 \leqslant t < 1; \\
\mathcal{P}_1 &= \{\{\omega_1, \omega_2\}, \{\omega_3, \omega_4\}\}, & 1 \leqslant t < 2; \\
\mathcal{P}_2 &= \{\{\omega_1\}, \{\omega_2\}, \{\omega_3\}, \{\omega_4\}\}, & 2 \leqslant t \leqslant T.
\end{aligned}$$

At time $t = 0$ we only know that some event $\omega \in \Omega$ will happen; at time $t = 2$ we will know which event $\omega^* \in \Omega$ has happened. At times $0 \leqslant t < 1$ we only know that $\omega^* \in \Omega$. At time points after $t = 1$ and strictly before $t = 2$, i.e. $1 \leqslant t < 2$, we know to which state the process has jumped at time $t = 1$: a or b. So at that time we will know to which set of $\mathcal{P}_1$ ω^* belongs: it will belong to $\{\omega_1, \omega_2\}$ if we jumped at time $t = 1$ to a, and to $\{\omega_3, \omega_4\}$ if we jumped to b. Finally, at time $t = 2$, we will know to which set of $\mathcal{P}_2$ ω^* will belong, in other words we will then know the complete path of the process.

During the flow of time we thus learn about the partitions. Having the information $\mathcal{F}_t$ revealed is equivalent to knowing in which set of the partition of that time the event ω^* is. The partitions become finer in each step and thus information on ω^* becomes more detailed.

2.2 Classes of Processes

2.2.1 Markov Processes

A *Markov process* is a particular type of stochastic process where only the present value of a variable is relevant for predicting the future. The past history of the variable and the way that the present has emerged from the past are irrelevant.

Stock prices are usually assumed to follow a Markov process to some degree. If the stock price follows a Markov process, our predictions of the future should be unaffected by the price one week ago, one month ago, or one year ago. The only relevant piece of information is the price now. Predictions are uncertain and must be expressed in terms of probability distributions. The Markov property implies that the probability distribution of the price at any particular future time is not dependent on the particular path followed by the price in the past.

If our stock-price process $S = \{S_t, 0 \leqslant t \leqslant T\}$ is Markovian and if we denote by $\mathbb{F} = \{\mathcal{F}_t, 0 \leqslant t \leqslant T\}$ the natural filtration of S (intuitively, $\mathcal{F}_t$ contains all our market information up to time t), then, with a little abuse of notation, we can write for a well-behaved function f:

$$E[f(S_T) \mid \mathcal{F}_t] = E[f(S_T) \mid S_t].$$

2.2.2 Martingales

A stochastic process $X = \{X_t, t \geqslant 0\}$ is a martingale relative to $(P, \mathbb{F})$ if

(i) X is $\mathbb{F}$-adapted,

(ii) $E[|X_t|] < \infty$ for all $t \geqslant 0$,

(iii) $E[X_t \mid \mathcal{F}_s] = X_s$, P-a.s. $(0 \leqslant s \leqslant t)$.

A martingale is 'constant on average', and models a fair game. This can be seen from the third condition: the best forecast of the unobserved future value X_t based on information at time s, $\mathcal{F}_s$, is the value X_s known at time s. In particular, the expected value of a martingale X at some time T (based on information at the initial time 0) equals its initial value X_0:

$$E[X_T \mid \mathcal{F}_0] = X_0.$$

2.2.3 Finite- and Infinite-Variation Processes

Consider a real-valued function $f : [a, b] \to \mathbb{R}$.

Càdlàg Function

Assume that for all $t \in (a, b]$ the function f is right continuous and has a left limit. We say the process is càdlàg, from the French 'continue à droite et limites à gauche'; the term RCLL (right continuous left limit) is sometimes also used. Clearly, any continuous function is càdlàg.

If f is càdlàg, we will denote the left limit at each point $t \in (a, b]$ as $f(t-) = \lim_{s \uparrow t} f(s)$. We stress that $f(t-) = f(t)$ if and only if f is continuous at t. The jump at t is denoted by

$$\Delta f(t) = f(t) - f(t-).$$

(In)finite-Variation Function

Let $\mathcal{P} = \{a = t_1 < t_2 < \cdots < t_{n+1} = b\}$ be a partition on the interval $[a, b] \subset \mathbb{R}$. We define the variation of the function f over the partition $\mathcal{P}$ by

$$\mathrm{var}_{\mathcal{P}}(f) = \sum_{i=1}^{n} |f(t_{i+1}) - f(t_i)|.$$

If the supremum over all partitions is finite, $\sup_{\mathcal{P}} \mathrm{var}_{\mathcal{P}}(f) < \infty$, we say that f has finite variation on $[a, b]$. If this is not the case, the function is said to be of infinite variation. If f is defined on $\mathbb{R}$ or on $[0, \infty)$, it is said to have finite variation if it has finite variation on each compact interval. Again, if this is not the case, the function is said to be of infinite variation.

Note that every nondecreasing f is of finite variation. Conversely, if f is of finite variation, then it can always be written as the difference of two nondecreasing functions.

Functions of finite variation are important in integration theory. More precisely, if we want to integrate over some interval a continuous function g with respect to an integrator f, then we are able to define the Stieltjes integral $\int_I g(x)\,\mathrm{d}f(x)$ as a limit of Riemann sums only if f has finite variation.

We say that a stochastic process $X = \{X_t, t \geqslant 0\}$ is of finite variation if the sample paths are of finite variation with probability 1. If this is not the case, we say that the process is of infinite variation. A typical example of a finite-variation process is the Poisson process (see Chapter 5). Note also that Brownian motion (see Chapter 3) is of infinite variation.

2.3 Characteristic Functions

The characteristic function ϕ of a distribution, or equivalently of a random variable X, is the Fourier–Stieltjes transform of the distribution function $F(x) = P(X \leqslant x)$:

$$\phi_X(u) = E[\exp(iuX)] = \int_{-\infty}^{+\infty} \exp(iux)\,\mathrm{d}F(x).$$

Some properties of characteristic functions are that $\phi(0) = 1$ and $|\phi(u)| \leqslant 1$, for all $u \in \mathbb{R}$. Moreover, the characteristic function always exists and is continuous. Most important is the fact that ϕ determines the distribution function F uniquely. The moments of X can also easily be derived from ϕ. Suppose X has a kth moment ($k \in \{0, 1, 2, \ldots\}$), i.e. assume $E[|X|^k] < \infty$, then

$$E[X^k] = i^{-k} \left. \frac{\mathrm{d}}{\mathrm{d}u^k}\phi(u) \right|_{u=0}.$$

Table 2.1   Functions related to the characteristic function.

Name	Definition
cumulant function	$k(u) = \log E[\exp(-uX)] = \log \phi(iu)$
moment-generating function	$\vartheta(u) = E[\exp(uX)] = \phi(-iu)$
cumulant characteristic function	$\psi(u) = \log E[\exp(iuX)] = \log \phi(u)$

Based on this we call $\vartheta(u) = \phi(-iu)$, when it exists for all $u \in \mathbb{R}$, the moment-generating function. We have

$$E[X^k] = \frac{\mathrm{d}}{\mathrm{d}u^k} \vartheta(u) \bigg|_{u=0}.$$

We also make frequent use of the function $k(u) = \log \phi(iu)$, which we call the cumulant function of X (see Table 2.1).

Finally, we note that if X and Y are two independent random variables with characteristic functions ϕ_X and ϕ_Y, respectively, then the characteristic function of $X + Y$ is given by $\phi_{X+Y}(u) = \phi_X(u)\phi_Y(u)$. In other words, characteristic functions take convolutions into multiplication. For a general introduction to characteristic functions, see Lukacs (1970).

2.4   Stochastic Integrals and SDEs

Stochastic integration was introduced by Itô in 1941, hence the name Itô calculus. It gives meaning to

$$\int_0^t X_u \, \mathrm{d}Y_u$$

for suitable stochastic processes $X = \{X_u, u \geqslant 0\}$ and $Y = \{Y_u, u \geqslant 0\}$, the integrand and the integrator. Because we will take as integrators processes of infinite (unbounded) variation on every interval (e.g. Brownian motion), the first thing to note is that stochastic integrals can be quite different from classical deterministic integrals. We take for granted Itô's fundamental insight that stochastic integrals can be defined for a suitable class of integrands.

As with any ordinary and partial differential equations (ODEs and PDEs) in a deterministic setting, the two most basic questions for stochastic differential equations (SDEs) are those of existence and uniqueness of solutions. To obtain existence and uniqueness results, we have to impose reasonable regularity conditions on the coefficients occurring in the differential equation. Naturally, SDEs contain all the complications of their nonstochastic counterparts, and more besides. See Protter (1990) for the general theory of stochastic integrals and SDEs.

The SDEs we encounter always have a unique solution and are of the following form:

$$\mathrm{d}X_t = a(t, X_t) \, \mathrm{d}t + b(t, X_t) \, \mathrm{d}Y_t, \quad X_0 = x_0. \tag{2.1}$$

The solution to such an SDE is a stochastic process $X = \{X_t, t \geqslant 0\}$, which satisfies

$$X_t = \int_0^t a(u, X_u)\, du + \int_0^t b(u, X_u)\, dY_u, \quad X_0 = x_0.$$

2.5 Financial Mathematics in Continuous Time

This section discusses the general principles of continuous-time modelling of financial markets.

2.5.1 Equivalent Martingale Measure

We say that a probability measure Q (defined on $(\Omega, \mathcal{F}_T)$) is an equivalent martingale measure if

- Q is equivalent to P, i.e. they have the same null sets (events which cannot happen under P also cannot happen under Q and vice versa);

- the discounted stock-price process $\tilde{S} = \{\tilde{S}_t = \exp(-rt)S_t, t \geqslant 0\}$ is a martingale under Q.

The existence of an equivalent martingale measure is related to the absence of arbitrage, while the uniqueness of the equivalent martingale measure is related to market completeness.

Existence of an Equivalent Martingale Measure

In discrete time and with finitely many states the existence of an equivalent martingale measure is equivalent to the absence of arbitrage, while the uniqueness of the equivalent martingale measure is equivalent to market completeness. In our continuous-time setting, existence of an equivalent martingale measure implies the absence of arbitrage, but the implication in the reverse direction is not valid. Essentially, the hypothesis of no-arbitrage is too weak to deduce the existence of an equivalent martingale measure. The strengthening required is that it should not be possible to construct an approximation to an arbitrage opportunity in some limiting sense, and then it does follow that there exists an equivalent martingale measure. The first results in this direction are due to Kreps (1981). The strongest results in this direction are due to Delbaen and Schachermayer (1994). They show that under the hypothesis that there is 'no free lunch with vanishing risk' (there is no random sequence of zero-cost trading strategies converging to a nonnegative, nonzero cash flow, with the random sequence bounded below by a negative constant), then there exists a martingale measure; the converse holds as well.

The existence of the equivalent martingale measure allows one to reduce the pricing of options on the risky asset to calculating the expected values of the discounted payoffs, not with respect to the physical (statistical) measure P, but with respect to the equivalent martingale measure Q (see Harrison and Kreps 1979; Harrison and Pliska 1981). We go into more detail in Section 2.5.2. If we work under Q, we often say that we are in a risk-neutral world, since under Q the expected return of the stock equals the risk-free return of the bank account:

$$E_Q[S_t \mid \mathcal{F}_0] = \exp(rt)S_0.$$

Uniqueness of Equivalent Martingale Measure

Besides pricing, an equally important problem is that of hedging. We say that a contingent claim can be perfectly hedged if there exists a (predictable) strategy which can replicate our claim in the sense that there is a dynamic portfolio, investing in the bank account and the stock, such that at every time point the value of the portfolio matches the value of the claim. The portfolio must be self-financing (we cannot subtract or pump in money). Moreover, in order to avoid problems that arise from the classical doubling strategy, the strategy must also be admissible, i.e. the portfolio's value must be bounded from below by a constant. The replicator's resources, while they can be huge, are nevertheless finite and bounded by a nonrandom constant. A market model is called complete if for every integrable contingent claim there exists an admissible self-financing strategy replicating the claim.

The question of completeness is linked with the uniqueness of the martingale measure, which is in turn linked with the mathematical predictable representation property (PRP) of a martingale. In probability theory a martingale M is said to have the PRP if, for any square-integrable random variable H ($\in \mathcal{F}_T$), we have

$$H = E[H] + \int_0^T a_s \, dM_s,$$

for some predictable process $a = \{a_s, 0 \leqslant s \leqslant T\}$. If we have such a representation, the predictable process a will give us our necessary self-financing admissible strategy. Unfortunately, the PRP is a rather delicate and exceptional property, which only a few martingales possess. Examples include Brownian motion, the compensated Poisson process, and the Azéma martingale (see Dritschel and Protter 1999).

The uniqueness of an equivalent martingale measure implies the PRP which in turn implies market completeness. However, there are examples where we have complete markets without uniqueness of the equivalent martingale measure (see Artzner and Heath 1995; Jarrow *et al.* 1999).

The PRP of Brownian motion leads to the completeness of the Black–Scholes model (see Chapter 3). Most models are not complete, and most practitioners believe the actual market is not complete. In incomplete markets, we have to choose an equivalent martingale measure in some way and this is not always clear. Actually, the market is choosing the martingale measure for us. The relation between the statistical measure (P) and the risk-neutral equivalent martingale measure (Q) is the subject of ongoing research.

2.5.2 Pricing Formulas for European Options

Given our market model, let $G(\{S_t, 0 \leqslant t \leqslant T\})$ denote the payoff of the derivative at its time of expiry T. In the case of the European call with strike price K, we have $G(\{S_t, 0 \leqslant t \leqslant T\}) = G(S_T) = (S_T - K)^+$. According to the fundamental theorem of asset pricing (see Delbaen and Schachermayer 1994), the arbitrage-free price V_t of the derivative at time $t \in [0, T]$ is given by

$$V_t = E_Q[\exp(-r(T - t))G(\{S_u, 0 \leqslant u \leqslant T\}) \mid \mathcal{F}_t],$$

where the expectation is taken with respect to an equivalent martingale measure Q and $\mathbb{F} = \{\mathcal{F}_t, 0 \leqslant t \leqslant T\}$ is the natural filtration of $S = \{S_t, 0 \leqslant t \leqslant T\}$. The factor $\exp(-r(T - t))$ is called the discounting factor.

Assume that we have selected our equivalent martingale measure Q, then we can compute option prices. If we know the density function of S_T, we can simply (numerically) calculate the price of a vanilla option as the discounted expected value of the payoff. On the other hand, we often do not have the density function available. However, in most cases we have the characteristic function of our stock-price process (or the logarithm of it) in the risk-neutral world at hand.

Let $C(K, T)$ be the price at time $t = 0$ of a European call option with strike K and maturity T. Next, we give an overview of some ways to calculate the option price.

Pricing Through the Density Function

If we know the density function, $f_Q(s, T)$, of our stock price at the expiry T under the risk-neutral measure Q, we can easily price European call and put options by simply calculating the expected value.

For a European call option with strike price K and time to expiration T, the value at time 0 is therefore given by the expectation of the payoff under the martingale measure:

$$
\begin{aligned}
C(K, T) &= E_Q[\exp(-rT) \max\{S_T - K, 0\}] \\
&= \exp(-rT) \int_0^\infty f_Q(s, T) \max\{s - K, 0\} \, ds \\
&= \exp(-rT) \int_K^\infty f_Q(s, T)(s - K) \, ds \\
&= \exp(-rT) \int_K^\infty f_Q(s, T) s \, ds - K \exp(-rT) \Pi_2,
\end{aligned}
$$

where Π_2 is the probability (under Q) of finishing in the money. Note that we have already assumed that f_Q lives on the nonnegative real numbers since the stock price is always greater than 0.

Pricing Through the Characteristic Function

Bakshi and Madan (2000) and Carr and Madan (1998) developed more explicit pricing methods for the classical vanilla options, which can be applied in general when the characteristic function of the risk-neutral stock-price process is known.

As usual let $S = \{S_t, 0 \leqslant t \leqslant T\}$ denote the stock-price process and denote by $\phi(u)$ the characteristic function of the random variable $\log S_T$, i.e.

$$\phi(u) = E[\exp(iu \log(S_T))].$$

Inversion of Distribution Function Transform.   Bakshi and Madan (2000) show very generally that we may write

$$C(K, T) = S_0 \Pi_1 - K \exp(-rT)\Pi_2, \tag{2.2}$$

where Π_1 and Π_2 are obtained by computing the integrals

$$\Pi_1 = \frac{1}{2} + \frac{1}{\pi} \int_0^\infty \mathrm{Re}\left(\frac{\exp(-iu \log K)E[\exp(i(u-i)\log S_T)]}{iu E[S_T]}\right) du$$

$$= \frac{1}{2} + \frac{1}{\pi} \int_0^\infty \mathrm{Re}\left(\frac{\exp(-iu \log K)\phi(u-i)}{iu\phi(-i)}\right) du,$$

$$\Pi_2 = \frac{1}{2} + \frac{1}{\pi} \int_0^\infty \mathrm{Re}\left(\frac{\exp(-iu \log K)E[\exp(iu \log S_T)]}{iu}\right) du$$

$$= \frac{1}{2} + \frac{1}{\pi} \int_0^\infty \mathrm{Re}\left(\frac{\exp(-iu \log K)\phi(u)}{iu}\right) du.$$

The probability of finishing in the money is Π_2. Similarly, the delta (i.e. the change in the value of the option compared with the change in the value of the underlying asset) of the option corresponds to Π_1.

Inversion of the Modified Call Price.   Let α be a positive constant such that the αth moment of the stock price exists. For all stock-price models encountered later on, a value of $\alpha = 0.75$ will typically do fine. Carr and Madan (1998) then showed that

$$C(K, T) = \frac{\exp(-\alpha \log(K))}{\pi} \int_0^{+\infty} \exp(-iv \log(K))\varrho(v)\, dv, \tag{2.3}$$

where

$$\varrho(v) = \frac{\exp(-rT)E[\exp(i(v-(\alpha+1)i)\log(S_T))]}{\alpha^2 + \alpha - v^2 + i(2\alpha+1)v} \tag{2.4}$$

$$= \frac{\exp(-rT)\phi(v-(\alpha+1)i)}{\alpha^2 + \alpha - v^2 + i(2\alpha+1)v}. \tag{2.5}$$

The fast Fourier transform can be used to invert the generalized Fourier transform of the call price. Put options can be priced using the put–call parity. This Fourier method was generalized to other types of options, such as power and self-quanto options, in Raible (2000).

2.6 Dividends

Up to now, we have assumed that the risky asset pays no dividends; but, in reality, stocks can sometimes pay some dividends to their holders. We assume that the amount and timing of the dividends during the life of an option can be predicted with certainty. Moreover, we will assume that the stock pays a continuous compound dividend yield at a rate q per annum. Other methods of paying dividends can be considered and techniques for dealing with this are described in the literature (see Hull 2000).

Continuous payment of a dividend yield at rate q means that our stock is following a process of the form,

$$S_t = \exp(-qt)\bar{S}_t,$$

where $\bar{S}$ describes the stock price's behaviour, not taking dividends into account. A stock which pays dividends continuously and an identical stock that does not pay dividends should provide the same overall return, i.e. dividends plus capital gains. The payment of dividends causes the growth of the stock price to be less than it would otherwise be by an amount q. In other words, if, with a continuous dividend yield of q, the stock price grows from S_0 to S_T at time T, then in the absence of dividends it would grow from S_0 to $\exp(qt)S_T$. Alternatively, in the absence of dividends it would grow from $\exp(-qt)S_0$ to S_T. This argument brings us to the fact that we get the same probability distribution for the stock price at time T in the following cases: (1) the stock starts at S_0 and pays a continuous dividend yield at rate q, and (2) the stock starts at price $\exp(-qt)S_0$ and pays no dividend yield.

The theory described up to now essentially still holds, although we need to take into account the above observations. For example, the put–call parity for a stock with dividend yield q can be obtained from the put–call parity for nondividend-paying stocks. With no dividends we obtained (with the same notation as in Chapter 1)

$$S_t + P_t - C_t = K \exp(-r(T - t)).$$

If we now take dividends into account, the change comes down to replacing S_t with $S_t \exp(-q(T - t))$. We have

$$\exp(-q(T - t))S_t + P_t - C_t = K \exp(-r(T - t)).$$

This relation can be proved by considering the portfolio consisting of $\exp(-q(T - t))$ stocks, 1 put option and -1 call option. We reinvest the dividends on the shares instantaneously in additional shares, i.e. at some future time point $t \leqslant s \leqslant T$ we have $\exp(-q(T - s))$ stocks; at the expiry date of the option we own 1 stock, 1 put and -1 call. The value of the portfolio at that time thus always equals K. By the no-arbitrage argument, the value of the portfolio at time T must equal $K \exp(-r(T - t))$, the value of a future payment (at time T) of K at time t.

The questions about existence and uniqueness of an equivalent martingale measure Q are slightly changed. It is now required that there exists (a unique) equivalent martingale measure making the discounted stock prices $\tilde{S}_t = \exp(-(r - q)t)S_t$ a martingale. We are thus discounting, not with factor r, but with factor $r - q$. If we require that $\tilde{S} = \{\tilde{S}_t, t \geqslant 0\}$ is a martingale under Q, then

$$E_Q[S_t] = \exp((r - q)t)S_0.$$

In the risk-neutral world the rate of return on the stock portfolio is the risk-free rate of return r: $r - q$ on the stock price, together with the dividends, which have a rate of return q.

If our asset is an index, the dividend yield is the (weighted) average of the dividend yields on the stocks comprising the index. In the case of our S&P 500 example, the dividend yield q assumed on the index on the day of our option dataset, i.e. on 18 April 2002, was taken to be 1.20%, whereas at that time the short rate r was 1.90%.

In practice, the dividend yield can be determined from the forward price of the asset. A forward contract is a very simple derivative. It is the agreement to buy or sell an asset at a certain future time for a certain price, the delivery price. At the time the contract is entered into, the delivery price is chosen so that the value of the forward is zero. This means that it costs nothing to buy or sell the contract. For an asset paying a continuous yield at rate q, the delivery price of a forward contract expiring at time T is given by $F = S_0 \exp((r - q)T)$. Assuming that the short rate r and the delivery price of the forward as given, q can easily be obtained.

3

The Black–Scholes Model

This chapter develops the most basic and well-known continuous-time, continuous-variable stochastic process for stock prices. An understanding of this process is the first step towards the understanding of the pricing of options in other more complicated markets.

3.1 The Normal Distribution

Definition

The Normal distribution, Normal(μ, σ^2), is one of the most important distributions, and is found in many areas of study. It lives on the real line, has mean $\mu \in \mathbb{R}$ and variance $\sigma^2 > 0$. Its characteristic function is given by

$$\phi_{\text{Normal}}(u; \mu, \sigma^2) = \exp(iu\mu)\exp(-\tfrac{1}{2}\sigma^2 u^2)$$

and the density function is

$$f_{\text{Normal}}(x; \mu, \sigma^2) = \frac{1}{\sqrt{2\pi\sigma^2}}\exp\left(-\frac{(x-\mu)^2}{2\sigma^2}\right).$$

Properties

The Normal(μ, σ^2) distribution is symmetric around its mean, and always has a kurtosis equal to 3:

	Normal(μ, σ^2)
mean	μ
variance	σ^2
skewness	0
kurtosis	3

Lévy Processes in Finance   W. Schoutens
© 2003 John Wiley & Sons, Ltd   ISBN: 0-470-85156-2

The Cumulative Probability Distribution Function: N(x)

We will denote by

$$N(x) = \int_{-\infty}^{x} f_{\text{Normal}}(u; 0, 1)\, du \tag{3.1}$$

the cumulative probability distribution function for a variable that is standard Normally distributed (Normal(0, 1)). This special function is built into most mathematical software packages. The next approximation produces values of $N(x)$ to within six decimal places of the true value (see Abramowitz and Stegun 1968):

$$N(x) = \begin{cases} 1 - \dfrac{\exp(-x^2/2)}{\sqrt{2\pi}}(a_1 k + a_2 k^2 + a_3 k^3 + a_4 k^4 + a_5 k^5) & \text{for } x \geqslant 0, \\ 1 - N(-x) & \text{for } x < 0, \end{cases}$$

where

$$k = (1 + 0.231\,641\,9x)^{-1},$$
$$a_1 = 0.319\,381\,530,$$
$$a_2 = -0.356\,563\,782,$$
$$a_3 = 1.781\,477\,937,$$
$$a_4 = -1.821\,255\,978,$$
$$a_5 = 1.330\,274\,429.$$

3.2   Brownian Motion

The big brother of the Normal distribution is Brownian motion. Brownian motion is the dynamic counterpart – where we work with evolution in time – of its static counterpart, the Normal distribution. Both arise from the central limit theorem. Intuitively, this tells us that the suitably normalized sum of many small independent random variables is approximately Normally distributed. These results explain the ubiquity of the Normal distribution in a static context. If we work in a dynamic setting, i.e. with stochastic processes, Brownian motion appears in the same manner.

The History of Brownian Motion

The history of Brownian motion dates back to 1828, when the Scottish botanist Robert Brown observed pollen particles in suspension under a microscope and observed that they were in constant irregular motion. By doing the same with particles of dust, he was able to rule out that the motion was due to the pollen particles being 'alive'.

In 1900 Bachelier considered Brownian motion as a possible model for stock market prices. Bachelier's model was his thesis. At that time the topic was not thought worthy of study.

In 1905 Einstein considered Brownian motion as a model of particles in suspension. He observed that, if the kinetic theory of fluids was right, then the molecules of water would move at random and so a small particle would receive a random number of impacts of random strength and from random directions in any short period of time. Such a bombardment would cause a sufficiently small particle to move in exactly the way described by Brown. Einstein also used it to estimate Avogadro's number.

In 1923 Norbert Wiener defined and constructed Brownian motion rigorously for the first time. The resulting stochastic process is often called the Wiener process in his honour.

It was with the work of Samuelson (1965) that Brownian motion reappeared as a modelling tool in finance.

3.2.1 Definition

A stochastic process $X = \{X_t, t \geq 0\}$ is a *standard Brownian motion* on some probability space $(\Omega, \mathcal{F}, P)$ if

(i) $X_0 = 0$ a.s.,

(ii) X has independent increments,

(iii) X has stationary increments,

(iv) $X_{t+s} - X_t$ is Normally distributed with mean 0 and variance $s > 0$: $X_{t+s} - X_t \sim$ Normal$(0, s)$.

We shall henceforth denote standard Brownian motion by $W = \{W_t, t \geq 0\}$ (W for Wiener). Note that the second item in the definition implies that Brownian motion is a Markov process. Moreover, Brownian motion is the basic example of a Lévy process (see Chapter 5).

In the above, we have defined Brownian motion without reference to a filtration. In what follows, unless otherwise specified, we will always work with the natural filtration $\mathbb{F} = \mathbb{F}^W = \{\mathcal{F}_t, 0 \leq t \leq T\}$ of W. We have that Brownian motion is adapted with respect to this filtration and that increments $W_{t+s} - W_t$ are independent of $\mathcal{F}_t$.

Random-Walk Approximation of Brownian Motion

No construction of Brownian motion is easy. We take the existence of Brownian motion for granted (for more details see Billingsley (1995)). To gain some intuition into its behaviour, it is useful to compare Brownian motion with a simple symmetric random walk on the integers. More precisely, let $X = \{X_i, i = 1, 2, \ldots\}$ be a series of independent and identically distributed random variables with $P(X_i = 1) = P(X_i = -1) = 1/2$. Define the simple symmetric random walk $Z = \{Z_n, n = 0, 1, 2, \ldots\}$ as $Z_0 = 0$ and $Z_n = \sum_{i=1}^{n} X_i$, $n = 1, 2, \ldots$. Rescale this random walk as $Y_k(t) = Z_{\lfloor kt \rfloor}/\sqrt{k}$, where $\lfloor x \rfloor$ is the integer part of x. Then from the Central Limit Theorem, $Y_k(t) \to W_t$ as $k \to \infty$, with convergence in distribution (or weak convergence).

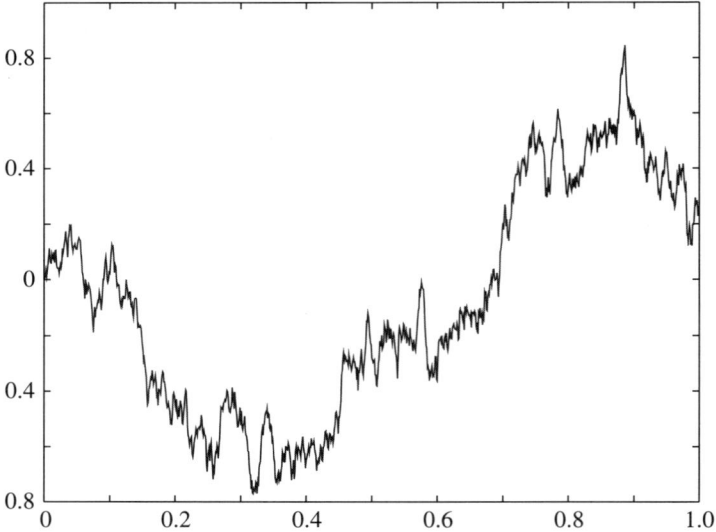

Figure 3.1   A sample path of a standard Brownian motion.

A realization of the standard Brownian motion is shown in Figure 3.1.

The random-walk approximation of the standard Brownian motion is shown in Figure 3.2. The process $Y_k = \{Y_k(t), t \geq 0\}$ is shown for $k = 1$ (i.e. the symmetric random walk), $k = 3$, $k = 10$ and $k = 50$. Clearly, we can see that $Y_k(t) \to W_t$.

3.2.2   Properties

Next, we look at some of the classical properties of Brownian motion.

Martingale Property

Brownian motion is one of the simplest examples of a martingale. We have, for all $0 \leq s \leq t$,

$$E[W_t \mid \mathcal{F}_s] = E[W_t \mid W_s] = W_s.$$

We also mention that we have

$$E[W_t W_s] = \min\{t, s\}.$$

Path Properties

We can prove that Brownian motion has continuous paths, i.e. W_t is a continuous function of t. However, the paths of Brownian motion are very erratic. They are, for example, nowhere differentiable. Moreover, we can also prove that the paths of Brownian motion are of infinite variation, i.e. their variation is infinite on every interval.

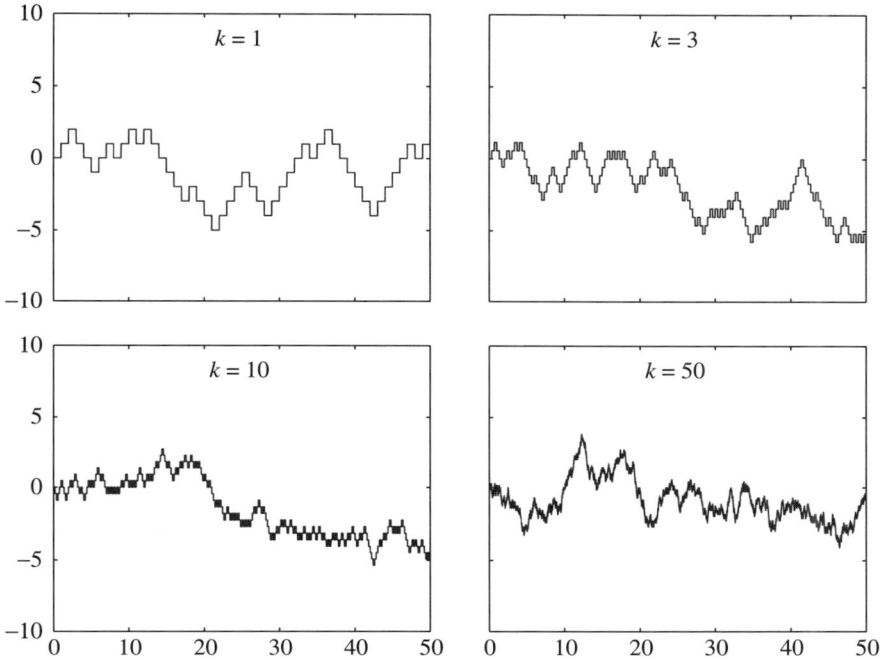

Figure 3.2   Random-walk approximation for standard Brownian motion.

Another property is that for a Brownian motion $W = \{W_t, t \geq 0\}$, we have that

$$P\left(\sup_{t \geq 0} W_t = +\infty \text{ and } \inf_{t \geq 0} W_t = -\infty\right) = 1.$$

This result tells us that the Brownian path will keep oscillating between positive and negative values.

Scaling Property

There is a well-known set of transformations of Brownian motion which produce another Brownian motion. One of these is the scaling property which says that if $W = \{W_t, t \geq 0\}$ is a Brownian motion, then, for every $c \neq 0$,

$$\tilde{W} = \{\tilde{W}_t = c W_{t/c^2}, t \geq 0\} \tag{3.2}$$

is also a Brownian motion.

3.3   Geometric Brownian Motion

Now that we have the Brownian motion W, we can introduce a stochastic process that is important for us, a relative of Brownian motion: *geometric Brownian motion*.

In the Black–Scholes model, the time evolution of a stock price $S = \{S_t, t \geq 0\}$ is modelled as follows. Consider how S will change in some small time interval from the present time t to a time $t + \Delta t$ in the near future. Writing ΔS_t for the change $S_{t+\Delta t} - S_t$, the return in this interval is $\Delta S_t / S_t$. It is economically reasonable to expect this return to decompose into two components, a *systematic* and a *random* part.

Let us first look at the systematic part. We assume that the stock's expected return over a period is proportional to the length of the period considered. This means that in a short interval of time $[S_t, S_{t+\Delta t}]$ of length Δt, the expected increase in S is given by $\mu S_t \Delta t$, where μ is some parameter representing the mean rate of the return of the stock. In other words, the deterministic part of the stock return is modelled by $\mu \Delta t$.

A stock price fluctuates stochastically, and a reasonable assumption is that the variance of the return over the interval of time $[S_t, S_{t+\Delta t}]$ is proportional to the length of the interval. So, the random part of the return is modelled by $\sigma \Delta W_t$, where ΔW_t represents the (Normally distributed) noise term (with variance Δt) driving the stock-price dynamics, and $\sigma > 0$ is the parameter that describes how much effect the noise has – how much the stock price fluctuates. In total, the variance of the return equals $\sigma^2 \Delta t$. Thus σ governs how volatile the price is, and is called the *volatility* of the stock. Putting this together, we have

$$\Delta S_t = S_t(\mu \Delta t + \sigma \Delta W_t), \quad S_0 > 0.$$

In the limit, as $\Delta t \to 0$, we have the stochastic differential equation:

$$\mathrm{d}S_t = S_t(\mu\,\mathrm{d}t + \sigma\,\mathrm{d}W_t), \quad S_0 > 0. \tag{3.3}$$

The above stochastic differential equation has the unique solution (see, for example, Bingham and Kiesel (1998) or Björk (1998))

$$S_t = S_0 \exp((\mu - \tfrac{1}{2}\sigma^2)t + \sigma W_t).$$

This (exponential) functional of Brownian motion is called geometric Brownian motion. Note that

$$\log S_t - \log S_0 = (\mu - \tfrac{1}{2}\sigma^2)t + \sigma W_t$$

has a Normal$(t(\mu - \tfrac{1}{2}\sigma^2), \sigma^2 t)$ distribution. Thus S_t itself has a *lognormal* distribution. This geometric Brownian motion model and the lognormal distribution which it entails form the basis for the Black–Scholes model for stock-price dynamics in continuous time.

In Figure 3.3, the realization of the geometric Brownian motion based on the sample path of the standard Brownian motion of Figure 3.1 is shown.

3.4   The Black–Scholes Option Pricing Model

In the early 1970s, Fischer Black, Myron Scholes and Robert Merton made a major breakthrough in the pricing of stock options by developing what has become known as

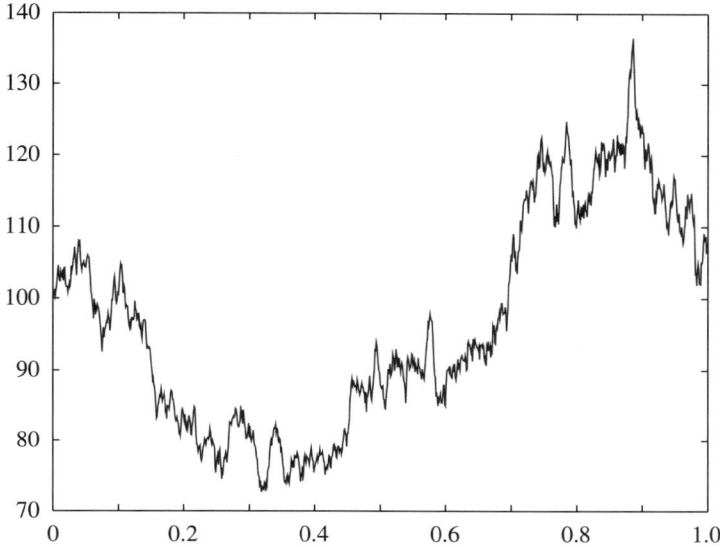

Figure 3.3   Sample path of geometric Brownian motion ($S_0 = 100$, $\mu = 0.05$, $\sigma = 0.40$).

the Black–Scholes model. The model has had huge influence on the way that traders price and hedge options. In 1997, the importance of the model was recognized when Myron Scholes and Robert Merton were awarded the Nobel Prize for economics. Sadly, Fischer Black died in 1995, otherwise he also would undoubtedly have been one of the recipients of this prize.

We show how the Black–Scholes model for valuing European call and put options on a stock works.

3.4.1   The Black–Scholes Market Model

Investors are allowed to trade continuously up to some fixed finite planning horizon T. The uncertainty is modelled by a filtered probability space $(\Omega, \mathcal{F}, P)$. We assume a frictionless market with two assets.

The first asset is one without risk (the bank account). Its price process is given by $B = \{B_t = \exp(rt), 0 \leqslant t \leqslant T\}$. The second asset is a risky asset, usually referred to as a stock, which pays a continuous dividend yield $q \geqslant 0$. The price process of this stock, $S = \{S_t, 0 \leqslant t \leqslant T\}$, is modelled by the geometric Brownian motion,

$$B_t = \exp(rt), \qquad S_t = S_0 \exp((\mu - \tfrac{1}{2}\sigma^2)t + \sigma W_t),$$

where $W = \{W_t, t \geqslant 0\}$ is a standard Brownian motion.

Note that, under P, W_t has a Normal$(0, t)$ and that $S = \{S_t, t \geqslant 0\}$ satisfies the SDE (3.3). The parameter μ reflects the drift and σ models the volatility; μ and σ are assumed to be constant over time.

We assume, as underlying filtration, the natural filtration $\mathbb{F} = (\mathcal{F}_t)$ generated by W. Consequently, the stock-price process $S = \{S_t, 0 \leqslant t \leqslant T\}$ follows a strictly positive adapted process. We call this market model the *Black–Scholes model*.

3.4.2 Market Completeness

As we saw in the previous chapter, questions of market completeness are related with the PRP. It was already known by Itô (1951) that Brownian motion possesses the PRP. The economic relevance of the representation theorem is that it shows that the Black–Scholes model is complete, that is, that every contingent claim can be replicated by a dynamic trading strategy. The desirable mathematical properties of Brownian motion are thus seen to have hidden within them desirable economic and financial consequences of real practical value (Bingham and Kiesel 1998).

3.4.3 The Risk-Neutral Setting

Since the Black–Scholes market model is complete, there exists only one equivalent martingale measure Q. It is not hard to see that under Q, the stock price is following a geometric Brownian motion again (Girsanov theorem). This risk-neutral stock-price process has the same volatility parameter σ, but the drift parameter μ is changed to the continuously compounded risk-free rate r minus the dividend yield q:

$$S_t = S_0 \exp((r - q - \tfrac{1}{2}\sigma^2)t + \sigma W_t).$$

Equivalently, we can say that under Q our stock-price process $S = \{S_t, 0 \leqslant t \leqslant T\}$ satisfies the following SDE:

$$dS_t = S_t((r - q)\, dt + \sigma\, dW_t), \quad S_0 > 0.$$

This SDE tells us that in a risk-neutral world the total return from the stock must be r; the dividends provide a return of q, the expected growth rate in the stock price, therefore, must be $r - q$.

Next, we will calculate European call option prices under this model.

3.4.4 The Pricing of Options under the Black–Scholes Model

General Pricing Formula

By the risk-neutral valuation principle, the price V_t at time t of a contingent claim with payoff function $G(\{S_u, 0 \leqslant u \leqslant T\})$ is given by

$$V_t = \exp(-(T - t)r) E_Q[G(\{S_u, 0 \leqslant u \leqslant T\}) \mid \mathcal{F}_t], \quad t \in [0, T]. \tag{3.4}$$

Furthermore, if the payoff function depends only on the time T value of the stock, i.e. $G(\{S_u, 0 \leqslant u \leqslant T\}) = G(S_T)$, then the above formula can be rewritten as (for simplicity, we set $t = 0$)

$$V_0 = \exp(-Tr)E_Q[G(S_T)]$$
$$= \exp(-Tr)E_Q[G(S_0 \exp((r - q - \tfrac{1}{2}\sigma^2)T + \sigma W_T))]$$
$$= \exp(-Tr) \int_{-\infty}^{+\infty} G(S_0 \exp((r - q - \tfrac{1}{2}\sigma^2)T + \sigma x)) f_{\text{Normal}}(x; 0, T)\, dx.$$

Black–Scholes PDE

Moreover, if $G(S_T)$ is a sufficiently integrable function, then the price is also given by $V_t = F(t, S_t)$, where F solves the *Black–Scholes partial differential equation*,

$$\frac{\partial}{\partial t}F(t, s) + (r - q)s\frac{\partial}{\partial s}F(t, s) + \tfrac{1}{2}\sigma^2 s^2\frac{\partial^2}{\partial s^2}F(t, s) - rF(t, s) = 0, \qquad (3.5)$$

$$F(T, s) = G(s).$$

This follows from the Feynman–Kac representation for Brownian motion (see, for example, Bingham and Kiesel 1998).

Explicit Formula for European Call and Put Options

Solving the Black–Scholes partial differential equation (3.5) is not always that easy. However, in some cases it is possible to evaluate explicitly the above expected value in the risk-neutral pricing formula (3.4).

Take, for example, a European call on the stock (with price process S) with strike K and maturity T (so $G(S_T) = (S_T - K)^+$). The Black–Scholes formulas for the price $C(K, T)$ at time zero of this European call option on the stock (with dividend yield q) is given by

$$C(K, T) = C = \exp(-qt)S_0 N(d_1) - K \exp(-rT)N(d_2),$$

where

$$d_1 = \frac{\log(S_0/K) + (r - q + \tfrac{1}{2}\sigma^2)T}{\sigma\sqrt{T}}, \qquad (3.6)$$

$$d_2 = \frac{\log(S_0/K) + (r - q - \tfrac{1}{2}\sigma^2)T}{\sigma\sqrt{T}} = d_1 - \sigma\sqrt{T}, \qquad (3.7)$$

and $N(x)$ is, as in (3.1), the cumulative probability distribution function for a variable that is standard Normally distributed (Normal$(0, 1)$).

From this, we can also easily (via the put–call parity) obtain the price $P(K, T)$ of the European put option on the same stock with same strike K and same maturity T:

$$P(K, T) = -\exp(-qt)S_0 N(-d_1) + K \exp(-rT)N(-d_2).$$

For the call, the probability (under Q) of finishing in the money corresponds to $N(d_2)$. Similarly, the delta (i.e. the change in the value of the option compared with the change in the value of the underlying asset) of the option corresponds to $N(d_1)$.

Note that the call option price is in the same form as the formula of Bakshi and Madan given in Equation (2.2).

4

Imperfections of the Black–Scholes Model

The Black–Scholes model has turned out to be very popular. One should bear in mind, however, that this elegant theory hinges on several crucial assumptions. We assumed that there was no market friction, such as taxes and transaction costs, and that there were no constraints on the stock holding, etc.

Moreover, empirical evidence suggests that the classical Black–Scholes model does not describe the statistical properties of financial time series very well. We will focus on two main problems. In Cont (2001) a more extended list of stylized features of financial data is given.

- We see that the log returns do not behave according to a Normal distribution.

- It has been observed that the volatilities or the parameters of uncertainty estimated (or more generally the environment) change stochastically over time and are clustered.

Next, we focus on these two problems in more detail.

4.1 The Non-Gaussian Character

4.1.1 Asymmetry and Excess Kurtosis

For a random variable X, we denote by

$$\mu_X = \mu = E[X]$$

its mean and by

$$\text{var}[X] = E[(X - \mu_X)^2] \geqslant 0$$

its variance. The square root of the variance $\sqrt{\text{var}[X]}$ is called the standard deviation. Recall that the standard deviation of a random variable following a Normal(μ, σ^2) distribution equals $\sigma > 0$.

Lévy Processes in Finance   W. Schoutens   .
© 2003 John Wiley & Sons, Ltd   ISBN: 0-470-85156-2

Table 4.1 Mean, standard deviation, skewness and kurtosis of major indices.

Index	Mean	SD	Skewness	Kurtosis
S&P 500 (1970–2001)	0.0003	0.0099	−1.6663	43.36
*S&P 500 (1970–2001)	0.0003	0.0095	−0.1099	7.17
S&P 500 (1997–1999)	0.0009	0.0119	−0.4409	6.94
Nasdaq-Composite	0.0015	0.0154	−0.5439	5.78
DAX	0.0012	0.0157	−0.4314	4.65
SMI	0.0009	0.0141	−0.3584	5.35
CAC-40	0.0013	0.0143	−0.2116	4.63

In Table 4.1 we summarize the empirical mean and the standard deviation for a set of popular indices. The first dataset (S&P 500 (1970–2001)) contains all daily log returns of the S&P 500 Index over the period 1970–2001. The second dataset (*S&P 500 (1970–2001)) contains the same data except for the exceptional log return (−0.2290) of the crash of 19 October 1987. All other datasets are over the period 1997–1999.

Next, we look at the empirical distribution of daily log returns of different indices. We will typically observe the asymmetry and fat tails of the empirical distribution.

Skewness

Skewness measures the degree to which a distribution is asymmetric. Skewness is defined to be the third moment about the mean, divided by the third power of the standard deviation:

$$\frac{E[(X - \mu_X)^3]}{\text{var}[X]^{3/2}}.$$

For a symmetric distribution (like the Normal(μ, σ^2)), the skewness is zero. If a distribution has a longer tail to the left than to the right, it is said to have negative skewness. If the reverse is true, then the distribution has a positive skewness.

If we look at the daily log returns of the different indices, we observe typically some significant (negative) skewness. In Table 4.1, we show the empirical skewness of the daily log returns for a set of popular indices. Recall that since the Normal distribution is symmetric it has a zero skewness.

Fat Tails and Excess Kurtosis

Next, we also show that large movements in asset price occur more frequently than in a model with Normal distributed increments. This feature is often referred to as *excess kurtosis* or *fat tails*; it is the main reason for considering asset price processes with jumps.

A way of measuring this fat tail behaviour is to look at the kurtosis, which is defined by

$$\frac{E[(X - \mu_X)^4]}{\text{var}[X]^2}.$$

For the Normal distribution (mesokurtic), the kurtosis is 3. If the distribution has a flatter top (platykurtic), the kurtosis is less than 3. If the distribution has a high peak (leptokurtic), the kurtosis is greater than 3.

In Table 4.1, we calculate the kurtosis of the daily log returns over the same periods for the same set of indices. We clearly see that our data always give rise to a kurtosis bigger than 3, indicating that the tails of the Normal distribution go to zero much faster than the empirical data suggest and that the empirical distribution is much more peaked than the Normal distribution. The fact that return distributions are more leptokurtic than the Normal has already been noted by Fama (1965).

4.1.2 Density Estimation

Finally, we look at the general picture of the empirical density and compare it with the Normal density.

Kernel Density Estimators

In order to estimate the empirical density, we make use of kernel density estimators. The goal of density estimation is to approximate the probability density function $f(x)$ of a random variable X. Assume that we have n independent observations $x_1, \ldots, x_n$ from the random variable X. The *kernel density estimator* $\hat{f}_h(x)$ for the estimation of the density $f(x)$ at point x is defined as

$$\hat{f}_h(x) = \frac{1}{nh} \sum_{i=1}^{n} K\left(\frac{x_i - x}{h}\right),$$

where $K(x)$ is a so-called kernel function and h is the bandwidth. We typically work with the so-called Gaussian kernel: $K(x) = \exp(-x^2/2)/\sqrt{2\pi}$. Other possible kernel functions are the so-called uniform, triangle, quadratic and cosinus kernel functions. In the above formula we also have to select the bandwidth h. With our Gaussian kernel, we use Silverman's rule-of-thumb value $h = 1.06\sigma n^{-1/5}$ (see Silverman 1986).

The Gaussian kernel density estimor based on the daily log returns of the S&P 500 Index over the period from 1970 until the end of 2001 is shown in Figure 4.1. We see a sharp peaked distribution. This tell us that, for most of the time, stock prices do not move that much; there is a considerable amount of mass around zero. Also plotted in Figure 4.1 is the Normal density with mean $\mu = 0.000\,311\,2$ and $\sigma = 0.0099$, corresponding to the empirical mean and standard deviation of the daily log returns.

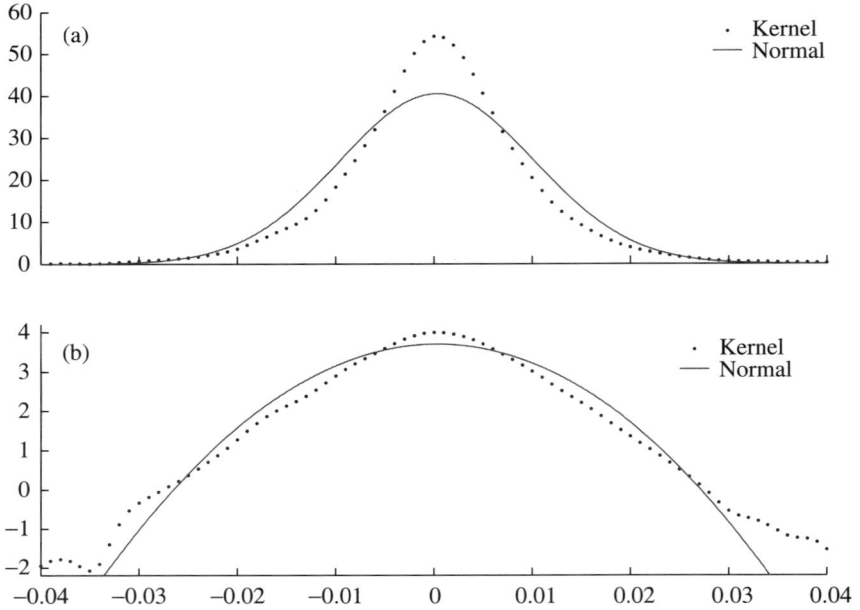

Figure 4.1 (a) Normal and Gaussian kernel density estimators and
(b) log densities of the daily log returns of the S&P 500 Index.

Semi-Heavy Tails

Density plots focus on the centre of the distribution; however, the tail behaviour is
also important. Therefore, we show in Figure 4.1 the log densities, i.e. $\log \hat{f}_h(x)$ and
the corresponding log of the Normal density. The log density of a Normal distribution
has a quadratic decay, whereas the empirical log density seems to have a much more
linear decay. This feature is typical for financial data and is often referred to as the
semi-heaviness of the tails. We say that a distribution or its density function $f(x)$ has
a semi-heavy tail if the tail of the density function behaves as

$$f(x) \sim \begin{cases} C_-|x|^{\rho_-} \exp(-\eta_-|x|) & \text{as } x \to -\infty, \\ C_+|x|^{\rho_+} \exp(-\eta_+|x|) & \text{as } x \to +\infty, \end{cases}$$

for some $\rho_-, \rho_+ \in \mathbb{R}$ and $C_-, C_+, \eta_-, \eta_+ \geqslant 0$.

In conclusion, we clearly see that the Normal distribution leads to a very bad fit.

4.1.3 Statistical Testing

Next, we will use some statistical tests to show that the Normal distribution does not
deliver a very good fit.

χ^2-Tests

A way of testing the goodness of fit is with the χ^2-test. The χ^2-test counts the number of sample points falling into certain intervals and compares them with the expected number under the null hypothesis.

More precisely, suppose we have n independent observations $x_1, \ldots, x_n$ from the random variable X and we want to test whether these observations follow a law with distribution D, depending on h parameters, which we all estimate by some method. First, make a partition $\mathcal{P} = \{A_1, \ldots A_m\}$ of the support (in our case $\mathbb{R}$) of D. The classes A_k can be chosen arbitrarily; we consider classes of equal width.

Let N_k, $k = 1, \ldots, m$, be the number of observations x_i falling into the set A_k; N_k/n is called the empirical frequency distribution. We will compare these numbers with the theoretical frequency distribution π_k, defined by

$$\pi_k = P(X \in A_k), \quad k = 1, \ldots, m,$$

through the Pearson statistic

$$\hat{\chi}^2 = \sum_{k=1}^{m} \frac{(N_k - n\pi_k)^2}{n\pi_k}.$$

If necessary, we collapse outer cells, so that the expected value $n\pi_k$ of the observations becomes always greater than 5.

We say a random variable χ_j^2 follows a χ^2-distribution with j degrees of freedom if it has a Gamma$(j/2, 1/2)$ law (see Chapter 5):

$$E[\exp(iu\chi_j^2)] = (1 - 2iu)^{-j/2}.$$

General theory says that the Pearson statistic $\hat{\chi}^2$ follows (asymptotically) a χ^2-distribution with $m - 1 - h$ degrees of freedom.

P-Value

The P-value of the $\hat{\chi}^2$ statistic is defined as

$$P = P(\chi_{m-1-h}^2 > \hat{\chi}^2).$$

In words, P is the probability that values are even more extreme (more in the tail) than our test statistic. It is clear that very small P-values lead to a rejection of the null hypothesis, because they are themselves extreme. P-values not close to zero indicate that the test statistic is not extreme and do not lead to a rejection of the hypothesis. To be precise, we reject the hypothesis if the P-value is less than our level of significance, which we take to be equal to 0.05.

Next, we calculate the P-value for the same set of indices. Table 4.2 shows the P-values of the test statistics. Similar tests can be found in, for example, Eberlein and Keller (1995).

We see that the Normal hypothesis is always rejected. Basically, we can conclude that a two-parameter model, such as the Normal one, is not sufficient to capture all

Table 4.2 Normal χ^2-test: P-values and class boundaries.

Index	P_{Normal}-value	Class boundaries	
S&P 500 (1970–2001)	0.0000	$-0.0300 + 0.0015i,$	$i = 0, \ldots, 40$
S&P 500 (1997–1999)	0.0421	$-0.0240 + 0.0020i,$	$i = 0, \ldots, 24$
DAX	0.0366	$-0.0225 + 0.0015i,$	$i = 0, \ldots, 30$
Nasdaq-Composite	0.0049	$-0.0300 + 0.0020i,$	$i = 0, \ldots, 30$
CAC-40	0.0285	$-0.0180 + 0.0012i,$	$i = 0, \ldots, 30$
SMI	0.0479	$-0.0180 + 0.0012i,$	$i = 0, \ldots, 30$

the features of the data. We need at least four parameters: a location parameter, a scale (volatility) parameter, an asymmetry (skewness) parameter and a (kurtosis) parameter describing the decay of the tails. We will see that the Lévy models introduced in the next chapter will have this required flexibility.

4.2 Stochastic Volatility

Another important feature missing from the Black–Scholes model is the fact that volatility or, more generally, the environment is changing stochastically over time.

Historical Volatility

It has been observed that the estimated volatilities (or, more generally, the parameters of uncertainty) change stochastically over time. This can be seen, for example, by looking at *historical volatilities*. Historical volatility is a retrospective measure of volatility. It reflects how volatile the asset has been in the recent past. Historical volatility can be calculated for any variable for which historical data are tracked.

For the S&P 500 Index, we estimated for every day from 1971 to 2001 the standard deviation of the daily log returns over a one-year period preceding the day. In Figure 4.2, for every day in the mentioned period, we plot the annualized standard deviation, i.e. we multiply the estimated standard deviation by the square root of the number of trading days in one calendar year. Typically, there are around 250 trading days in one year. This annualized standard deviation is called *the historical volatility*. Clearly, we see fluctuations in this historical volatility. Moreover, we see a kind of mean-reversion effect. The peak in the middle of the figure comes from the stock market crash on 19 October 1987; one-year windows including this day (with an extremal down-move) give rise to very high volatilities.

Volatility Clusters

Moreover, there is evidence for *volatility clusters*, i.e. there seems to be a succession of periods with high return variance and with low return variance. This can be seen, for example, in Figure 4.3, where the absolute log returns of the S&P 500 Index over

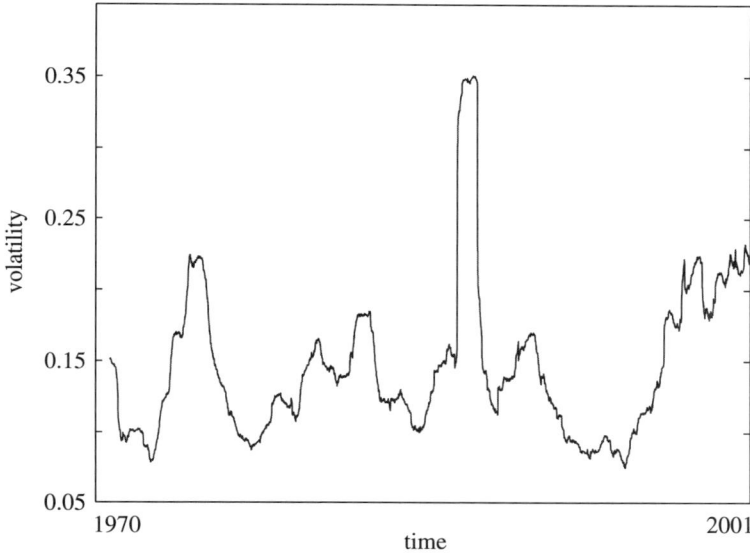

Figure 4.2   Historical volatility (one-year window) on S&P 500 (1970–2001).

a period of more than 30 years are plotted. We clearly see that there are periods with high absolute log returns and periods with lower absolute log returns. Large price variations are more likely to be followed by large price variations.

These observations motivate the introduction of models for asset price processes where volatility is itself stochastic.

4.3   Inconsistency with Market Option Prices

Calibration of Market Prices

If we estimate the model parameters by minimizing the root-mean-square error between market prices and the Black–Scholes model prices, we can observe an enormous difference. This can be seen in Figure 4.4 for the S&P 500 options. The volatility parameter which gives the best fit in the least-squared sense for the Black–Scholes model is $\sigma = 0.1812$ (in terms of years). Recall that the circles are market prices; the plus signs are the calibrated model prices.

Table 4.3 gives the relevant measures of fit we introduced in Chapter 1.

Implied Volatility

Another way to see that the classical Black–Scholes model does not correspond with option prices in the market is to look at the implied volatilities coming from the option prices. For every European call option with strike K and time to maturity T, we calculate the only (free) parameter involved, the volatility $\sigma = \sigma(K, T)$, so that

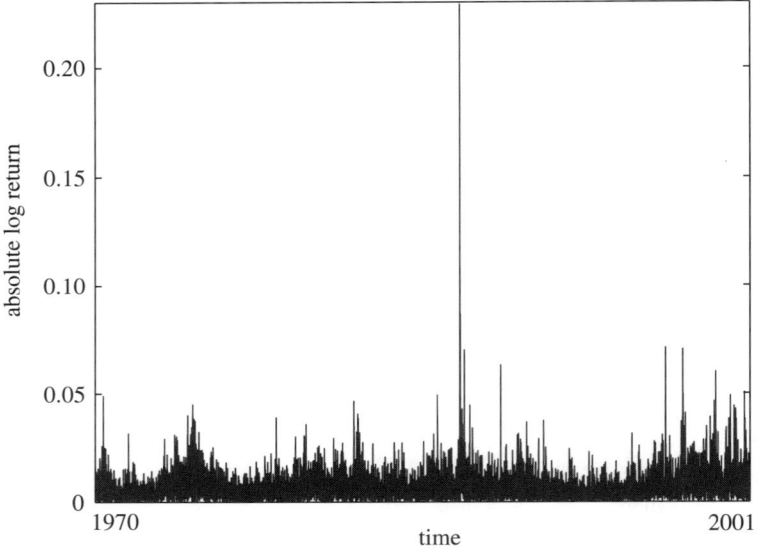

Figure 4.3 Volatility clusters: absolute log returns of the S&P 500 Index
between 1970 and 2001.

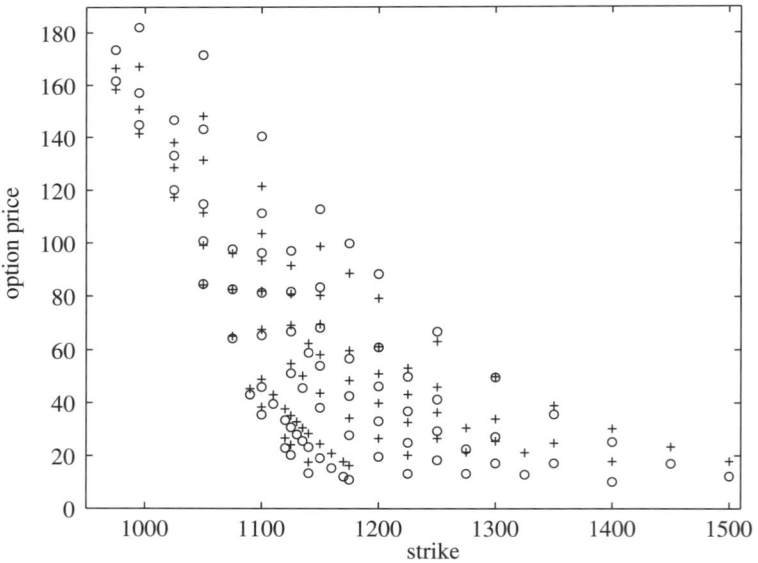

Figure 4.4 Black–Scholes ($\sigma = 0.1812$) calibration of S&P 500 options
(circles are market prices, pluses are model prices).

Table 4.3 APE, AAE and RMSE of the Black–Scholes model calibration of market option prices.

Model	APE	AAE	RMSE	ARPE
Black–Scholes	8.87%	5.4868	6.7335	16.92%

the theoretical option price (under the Black–Scholes model) matches the empirical one. This $\sigma = \sigma(K, T)$ is called the *implied volatility* of the option. Implied volatility is a timely measure: it reflects the market's perceptions today.

There is no closed formula to extract the implied volatility out of the call option price. We have to rely on numerical methods. One method of finding numerically implied volatilities is the classical Newton–Raphson iteration procedure. Denote by $C(\sigma)$ the price of the relevant call option as a function of volatility. If C is the market price of this option, we need to solve the transcendental equation

$$C = C(\sigma) \tag{4.1}$$

for σ. We start with some initial value we propose for σ; we denote this starting value by σ_0. In terms of years, it turns out that a σ_0 around 0.20 performs very well for most common stocks and indices. In general, if we denote by σ_n the value obtained after n iteration steps, the next value σ_{n+1} is given by

$$\sigma_{n+1} = \sigma_n - \frac{C(\sigma_n) - C}{C'(\sigma_n)},$$

where in the denominator C' refers to the differential with respect to σ of the call price function (this quantity is also referred to as the vega). For the European call option (under Black–Scholes) we have

$$C'(\sigma_n) = S_0\sqrt{T}N(d_1) = S_0\sqrt{T}N\left(\frac{\log(S_0/K) + (r - q + \frac{1}{2}\sigma_n^2)T}{\sigma_n\sqrt{T}}\right),$$

where S_0 is the current stock price, d_1 is as in (3.6) and $N(x)$ is the cumulative probability distribution of a Normal$(0, 1)$ random variable as in (3.1).

Next, we bring together for every maturity and strike this volatility σ in Figure 4.5, where the so-called volatility surface is shown. Under the Black–Scholes model, all σs should be the same; clearly, we observe that there is a huge variation in this volatility parameter both in strike and in time to maturity. We often say there is a volatility smile or skew effect. Again, this points up the fact that the Black–Scholes model is not appropriate and that traders already account for this deficiency in their prices.

Implied Volatility Models

Great care has to be taken over using implied volatilities to price options. Fundamentally, using implied volatilities is wrong. Taking different volatilities for different options on the same underlying asset gives rise to different stochastic models for one

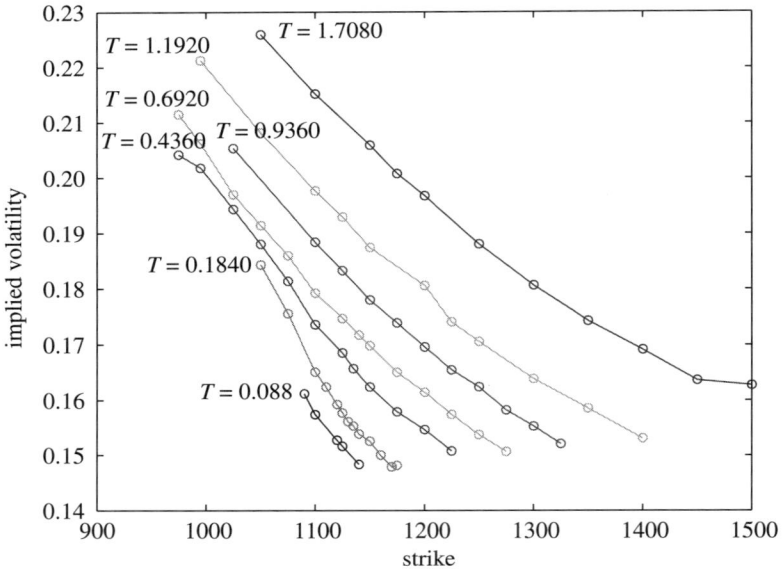

Figure 4.5   Implied volatilities.

asset. Moreover, the situation worsens in the case of exotic options. Shaw (1998) showed that if one tries to find the implied volatilities coming out of exotic options such as barrier options (see Chapter 9), there are cases where there are two or even three solutions to the implied volatility equation (for the European call option, see Equation (4.1)). Implied volatilities are thus not unique in these situations. More extremely, if we consider an up-and-out put barrier option, where the strike coincides with the barrier and the risk-free rate equals the dividend yield, the Black–Scholes price (for which a formula in closed form is available) is independent of the volatility. So, if the market price happens to coincide with the computed value, you can have any implied volatility you want. Otherwise, there is no implied volatility.

From this, it should be clear that great caution has to be taken by using European call option implied volatilities for exotic options with apparently similar characteristics (such as, for example, the same strike price). There is no guarantee that prices so obtained reflect true prices.

5

Lévy Processes and OU Processes

To price and hedge derivative securities, it is crucial to have a good model of the probability distribution of the underlying product. The most famous continuous-time model is the celebrated Black–Scholes model (see Chapter 3), which uses the Normal distribution to fit the log returns of the underlying.

As we have seen in the previous chapter, one of the main problems with the Black–Scholes model is that the data suggest that the log returns of stocks/indices are not Normally distributed as in the Black–Scholes model. The log returns of most financial assets do not follow a Normal law. They are skewed and have an actual kurtosis higher than that of the Normal distribution. Other more flexible distributions are needed. Moreover, not only do we need a more flexible static distribution, but in order to model the behaviour through time we need more flexible stochastic processes (which generalize Brownian motion).

Looking at the definition of Brownian motion, we would like to have a similar, i.e. with independent and stationary increments, process, based on a more general distribution than the Normal. However, in order to define such a stochastic process with independent and stationary increments, the distribution has to be infinitely divisible. Such processes are called Lévy processes, in honour of Paul Lévy, the pioneer of the theory.

To be useful in finance, the infinitely divisible distributions need to be able to represent skewness and excess kurtosis. In the late 1980s and in the 1990s, models having these characteristics were proposed for modelling financial data. The underlying Normal distribution was replaced by a more sophisticated infinitely divisible one.

Examples of such distributions, which can take into account skewness and excess kurtosis, are the Variance Gamma (VG), the Normal Inverse Gaussian (NIG), the CGMY (named after Carr, Geman, Madan and Yor), the (Generalized) Hyperbolic Model and the Meixner distributions. Madan and Seneta (1987, 1990) have proposed a Lévy process with VG distributed increments. The Hyperbolic Model was proposed by Eberlein and Keller (1995). In the same year, Barndorff-Nielsen (1995) proposed

Lévy Processes in Finance   W. Schoutens
© 2003 John Wiley & Sons, Ltd   ISBN: 0-470-85156-2

the NIG Lévy process. All three above-mentioned models were brought together as special cases of the Generalized Hyperbolic Model, which was developed by Eberlein and co-workers in a series of papers (see Eberlein and Prause 1998; Eberlein *et al.* 1998; Prause 1999). Recently, the CGMY model was introduced by Carr *et al.* (2002) and the Meixner model was used in Schoutens (2001).

In this chapter we give the theoretical background of Lévy processes and OU processes driven by Lévy processes. We give the definition of the most popular Lévy processes in use. Besides the processes used in finance we also look at the simpler class of subordinators on which the above-mentioned processes are built.

In the next chapter, we will show that Lévy models give a much better fit to the data and lead to a significant improvement with respect to the Black–Scholes model. Moreover, they can explain, at least in part, the so-called volatility smile.

In Chapter 7, we will incorporate stochastic volatility in these Lévy models. The stochastic behaviour for the volatility can be modelled using OU processes driven by Lévy processes. In this chapter, we have a close look at the special cases of OU processes which we later use for building the stochastic volatility models.

5.1 Lévy Processes

In this section Lévy processes are defined. General reference works on Lévy processes are by Bertoin (1996), Sato (1999) and Applebaum (2003). In Section 5.3 we look at a series of popular examples.

5.1.1 Definition

Suppose $\phi(u)$ is the characteristic function (see Section 2.3) of a distribution. If, for every positive integer n, $\phi(u)$ is also the nth power of a characteristic function, we say that the distribution is *infinitely divisible*.

We can define for every such infinitely divisible distribution a stochastic process, $X = \{X_t, t \geqslant 0\}$, called a Lévy process, which starts at zero and has independent and stationary increments such that the distribution of an increment over $[s, s + t]$, $s, t \geqslant 0$, i.e. $X_{t+s} - X_s$, has $(\phi(u))^t$ as its characteristic function.

Every Lévy process has a càdlàg (see Section 2.2.3) modification which is itself a Lévy process. We always work with this càdlàg version of the process. So, sample paths of a Lévy process are almost surely continuous from the right and have limits from the left. Moreover, we will always work in the sequel with the natural filtration generated by the Lévy process X.

The cumulant characteristic function $\psi(u) = \log \phi(u)$ is often called the *characteristic exponent*, which satisfies the following *Lévy–Khintchine formula*,

$$\psi(u) = i\gamma u - \tfrac{1}{2}\sigma^2 u^2 + \int_{-\infty}^{+\infty} (\exp(iux) - 1 - iux 1_{\{|x|<1\}}) \nu(dx), \qquad (5.1)$$

where $\gamma \in \mathbb{R}$, $\sigma^2 \geqslant 0$ and ν is a measure on $\mathbb{R}\backslash\{0\}$ with

$$\int_{-\infty}^{+\infty} \inf\{1, x^2\}\nu(\mathrm{d}x) = \int_{-\infty}^{+\infty} (1 \wedge x^2)\nu(\mathrm{d}x) < \infty.$$

We say that our infinitely divisible distribution has a triplet of Lévy characteristics (or Lévy triplet for short) $[\gamma, \sigma^2, \nu(\mathrm{d}x)]$. The measure ν is called the *Lévy measure* of X.

If the Lévy measure is of the form $\nu(\mathrm{d}x) = u(x)\,\mathrm{d}x$, we call $u(x)$ the *Lévy density*. The Lévy density has the same mathematical requirements as a probability density, except that it does not need to be integrable and must have zero mass at the origin.

From the Lévy–Khintchine formula, we see that, in general, a Lévy process consists of three independent parts: a linear deterministic part, a Brownian part and a pure jump part. The Lévy measure $\nu(\mathrm{d}x)$ dictates how the jumps occur. Jumps of sizes in the set A occur according to a Poisson process (see Section 5.3.1) with intensity parameter $\int_A \nu(\mathrm{d}x)$.

A subordinator is a nonnegative nondecreasing Lévy process. It is not hard to see that as such a subordinator has no Brownian part ($\sigma^2 = 0$), a nonnegative drift and a Lévy measure which is zero on the negative half-line (it has only positive increments). Note that a subordinator is nondecreasing and always of finite variation.

5.1.2 Properties

Path Properties

If $\sigma^2 = 0$ and $\int_{-1}^{+1} |x|\nu(\mathrm{d}x) < \infty$, it follows from standard Lévy process theory that the process is of finite variation. In that case the characteristic exponent can be re-expressed as

$$\psi(u) = \mathrm{i}\gamma' u + \int_{-\infty}^{+\infty} (\exp(\mathrm{i}ux) - 1)\nu(\mathrm{d}x)$$

for some γ', which we call the *drift coefficient*. In the finite-variation case, we can decompose the process into the difference of two increasing processes.

If $\sigma^2 = 0$ and $\int_{-1}^{+1} \nu(\mathrm{d}x) < \infty$, there are finitely many jumps in any finite interval. We say the process is of finite activity.

Because the Brownian motion is of infinite variation, a Lévy process with a Brownian component is of infinite variation. A pure jump Lévy process, i.e. one with no Brownian component ($\sigma^2 = 0$), is of infinite variation if and only if $\int_{-1}^{+1} |x|\nu(\mathrm{d}x) = \infty$. In that case special attention has to be paid to the small jumps. Basically, the sum of all jumps smaller than some $\epsilon > 0$ does not converge. However, the sum of the jumps compensated by their mean does converge. This peculiarity leads to the necessity of the compensator term $\mathrm{i}ux1_{\{|x|<1\}}$ in (5.1).

Predictable Representation Property

We denote the jump that a process $X = \{X_t, t \geq 0\}$ makes at time t by

$$\Delta X_t = X_t - X_{t-}.$$

Under some weak moment assumptions it was proved in Nualart and Schoutens (2000) that a Lévy process $X = \{X_t, 0 \leq t \leq T\}$ possesses a version of the predictable representation property (PRP). Every square integral random variable $F \in \mathcal{F}_T$ has a representation of the form,

$$F = E[F] + \sum_{i=1}^{\infty} \int_0^T a_s^{(i)} \mathrm{d}(H_s^{(i)} - E[H_s^{(i)}]), \tag{5.2}$$

where $a^{(i)} = \{a_s^{(i)}, 0 \leq s \leq T\}$ is predictable and $H^{(i)} = \{H_s^{(i)}, 0 \leq s \leq T\}$ is the power jump process of order i, i.e. $H_s^{(1)} = X_s$ and

$$H_s^{(i)} = \sum_{0 < u \leq s} (\Delta X_u)^i, \quad i = 2, 3, \ldots.$$

Since Brownian motion $W = \{W_t, t \geq 0\}$ has continuous paths, there are no jumps and $H_s^{(i)} = 0$ for $i \geq 2$. In this case the infinite sum reduces to only one term and we obtain that Brownian motion possesses the classical PRP,

$$F = E[F] + \int_0^T a_s \, \mathrm{d}W_s,$$

where $a = a^{(1)}$ is predictable. Note that this implies the completeness of the Black–Scholes model. It is only in the Poisson case (see Section 5.3.1) that a similar simplification to one term can be made. Therefore, the more realistic market models, based on a non-Brownian and non-Poissonian Lévy process, will lead to incomplete market models.

Paul Lévy (1886–1971)

The name Lévy process refers to one of the greatest mathematicians of the 20th century: Paul Lévy.

Paul Lévy was born in Paris in 1886. He studied at the École Polytechnique, obtained a doctoral degree in mathematics from the University of Paris and became professor at the École Polytechnique in 1913. He became one of the pioneers of modern probability theory, which was at that time in its early stages. He made important discoveries in the theory of stochastic processes. He proved the Central Limit Theorem using characteristic functions, independently from Lindeberg, who used convolution techniques. He studied various properties of Brownian motion and discovered the class of stable distributions. His main books are *Leçons d'analyse fonctionnelle* (1922), *Calcul des probabilités* (1925), *Théorie de l'addition des variables aléatoires* (1937–1954) and *Processus stochastiques et mouvement brownien* (1948).

During World War I Lévy served in the artillery and was involved in using his mathematical skills in solving problems concerning defence against attacks from the air. In 1963, he was elected to honorary membership of the London Mathematical Society. In the following year he was elected to the Académie des Sciences. Paul Lévy died in Paris on 15 December 1971.

5.2 OU Processes

In this section we give a brief introduction to self-decomposability and Ornstein–Uhlenbeck (OU) processes (driven by Lévy processes), which were introduced by Barndorff-Nielsen and Shephard (2001a,b, 2003b) as a model to describe volatility in finance. Further work on OU processes can be found in Wolfe (1982), Sato and Yamazato (1982), Jurek and Vervaat (1983) and Sato et al. (1994). For some notes about the origin, see Bingham (1998).

5.2.1 Self-Decomposability

Let ϕ be the characteristic function of a random variable X. Then X is self-decomposable if

$$\phi(u) = \phi(cu)\phi_c(u)$$

for all $u \in \mathbb{R}$ and all $c \in (0, 1)$ and for some family of characteristic functions $\{\phi_c : c \in (0, 1)\}$. It is also said that in that case the law of X belongs to Lévy's class L. A random variable with law in L is infinitely divisible.

A further important characterization of the class L as a subclass of the set of all infinitely divisible distributions in terms of the Lévy measure is the following equivalence.

Let $v(\mathrm{d}x)$ denote the Lévy measure of an infinitely divisible measure P on $\mathbb{R}$. Then the following statements are equivalent.

(1) P is self-decomposable.

(2) The functions (in s) on the positive half-line given by $v((-\infty, -e^s])$ and $v([e^s, \infty))$ are both convex.

(3) $v(\mathrm{d}x)$ is of the form $v(\mathrm{d}x) = u(x)\,\mathrm{d}x$ with $|x|u(x)$ increasing on $(-\infty, 0)$ and decreasing on $(0, \infty)$.

If u is differentiable, then the necessary and sufficient condition (2) may be re-expressed as

$$u(x) + xu'(x) \leqslant 0, \quad \text{for } x \neq 0. \tag{5.3}$$

The equivalence of (1), (2) and (3) is due to Lévy (1937). A proof may be found also in Bar-lev et al. (1992); see also Sato (1999).

5.2.2   OU Processes

Definition

An important role will also be played by processes driven by a Lévy process. We consider processes which are defined by the following SDE,

$$dy_t = -\lambda y_t \, dt + dz_{\lambda t}, \quad y_0 > 0, \tag{5.4}$$

where the process z_t is a subordinator; more precisely, it is a Lévy process with no Brownian part, nonnegative drift and only positive increments. We will call these processes $\{y_t, t \geq 0\}$ Ornstein–Uhlenbeck (OU) processes. The rate parameter λ is arbitrary positive and $z = \{z_t, t \geq 0\}$ is called the Background Driving Lévy Process (BDLP).

As z is an increasing process and $y_0 > 0$, it is clear that the process y is strictly positive. Moreover, it is bounded from below by the deterministic function $y_0 \exp(-\lambda t)$.

Remark 5.1. OU processes based on a general Lévy process, not necessarily a subordinator, can also be defined. However, for our analysis we will only need the special case considered above.

Remark 5.2. We can simply include a drift term in the SDE (5.4). In fact, a stochastic differential equation,

$$dy_t = (\alpha - \lambda y_t) \, dt + dz_{\lambda t}, \quad y_0 > 0,$$

can be recast in the form of (5.4) simply by defining a new BDLP $\tilde{z} = \{\tilde{z}, t \geq 0\}$ by

$$\tilde{z}_t = z_t + \lambda^{-1} \alpha t.$$

D-OU Processes and OU-D Processes

The process $y = \{y_t, t \geq 0\}$ is strictly stationary on the positive half-line, i.e. there exists a law D, called the stationary law or the marginal law, such that y_t will follow the law D for every t if the initial y_0 is chosen according to D. The process y moves up entirely by jumps and then tails off exponentially. In Barndorff-Nielsen and Shephard (2001a) some stochastic properties of y are studied. They established the notation that if y is an OU process with marginal law D, then we say that y is a D-OU process. Further, if the BDLP at time $t = 1$, i.e. z_1, has law $\tilde{D}$, then we say y is an OU-$\tilde{D}$ process.

In essence, given a one-dimensional distribution D (not necessarily restricted to the positive half-line), there exists a (stationary) OU process whose marginal law is D (i.e. a D-OU process) if and only if D is self-decomposable. We have by standard results (see Barndorff-Nielsen and Shephard 2001a) that

$$y_t = \exp(-\lambda t) y_0 + \int_0^t \exp(-\lambda(t-s)) \, dz_{\lambda s}$$

$$= \exp(-\lambda t) y_0 + \exp(-\lambda t) \int_0^{\lambda t} \exp(s) \, dz_s.$$

In the case of a D-OU process, let us denote by $k_D(u)$ the cumulant function of the self-decomposable law D and by $k_z(u)$ the cumulant function of the BDLP at time $t = 1$, i.e. $k_z(u) = \log E[\exp(-uz_1)]$; then both are related through the formula (see, for example, Barndorff-Nielsen 2001):

$$k_z(u) = u\frac{dk_D(u)}{du}.$$

Let us denote the Lévy measure of z_1 (the BDLP at time $t = 1$) by $W(dx)$. If the Lévy density u of the self-decomposable law D is differentiable, then the Lévy measure W has a density w, and u and w are related by

$$w(x) = -u(x) - xu'(x). \tag{5.5}$$

Tail Mass Function

Let the tail mass function of $W(dx)$ be

$$W^+(x) = \int_x^\infty w(y)\,dy,$$

we have from Barndorff-Nielsen (1998)

$$W^+(x) = xu(x).$$

Finally, we shall denote the inverse function of W^+ by W^{-1}, i.e.

$$W^{-1}(x) = \inf\{y > 0 : W^+(y) \leqslant x\}.$$

Integrated OU Process

An important related process will be the integral of y_t. Barndorff-Nielsen and Shephard called this the integrated OU process (intOU); we will denote this process by $Y = \{Y_t, t \geqslant 0\}$:

$$Y_t = \int_0^t y_s\,ds.$$

A major feature of the intOU process Y is

$$Y_t = \lambda^{-1}(z_{\lambda t} - y_t + y_0)$$
$$= \lambda^{-1}(1 - \exp(-\lambda t))y_0 + \lambda^{-1}\int_0^t (1 - \exp(-\lambda(t - s)))\,dz_{\lambda s}. \tag{5.6}$$

An interesting characteristic is that $Y = \{Y_t, t \geqslant 0\}$ has continuous sample paths when $\lambda > 0$, while $z = \{z_t, t \geqslant 0\}$ and $y = \{y_t, t \geqslant 0\}$ have jumps.

We can show (see Barndorff-Nielsen and Shephard 2001a) that, given y_0,

$$\log E[\exp(iuY_t) \mid y_0]$$
$$= \lambda \int_0^t k(u\lambda^{-1}(1 - \exp(-\lambda(t - s)))) \, ds + iuy_0\lambda^{-1}(1 - \exp(-\lambda t)),$$

where $k(u) = k_z(u) = \log E[\exp(-uz_1)]$ is the cumulant function of z_1. From this we obtain that, for every $t \geqslant 0$,

$$E[Y_t \mid y_0] = \lambda^{-1}(1 - \exp(-\lambda t))y_0 + \lambda^{-1}E[z_1](\lambda t - 1 - \exp(-\lambda t)),$$
$$\text{var}[Y_t \mid y_0] = \lambda^{-2}\text{var}[z_1](\lambda t - 2 + 2\exp(-\lambda t) + \tfrac{1}{2} - \tfrac{1}{2}\exp(-2\lambda t)).$$

5.3   Examples of Lévy Processes

In the following sections we list a number of popular Lévy processes. We start with some so-called subordinators. Next, we look at processes that live on the real line. We pay attention to their density function, their characteristic function, their Lévy triplets, together with some of their properties. We compute moments, variance, skewness and kurtosis, if possible, and look at the semi-heaviness of the tails.

5.3.1   The Poisson Process

Definition

The Poisson process is the simplest Lévy process we can think of. It is based on the Poisson(λ), $\lambda > 0$, distribution, which has

$$\phi_{\text{Poisson}}(u; \lambda) = \exp(\lambda(\exp(iu) - 1))$$

as characteristic function. The Poisson distribution lives on the nonnegative integers $\{0, 1, 2, \ldots\}$; the probability mass at point j equals

$$\exp(-\lambda)\frac{\lambda^j}{j!}.$$

Since the Poisson(λ) distribution is infinitely divisible, we can define a Poisson process $N = \{N_t, t \geqslant 0\}$ with intensity parameter $\lambda > 0$ as the process which starts at zero, has independent and stationary increments and where the increment over a time interval of length $s > 0$ follows a Poisson(λs) distribution. The Poisson process turns out to be an increasing pure jump process, with jump sizes always equal to 1. This means that the Lévy triplet is given by $[0, 0, \lambda\delta(1)]$, where $\delta(1)$ denotes the Dirac measure at point 1, i.e. a measure with a mass of only 1 at point 1. The time between two consecutive jumps follows an exponential distribution with mean λ^{-1}, i.e. a Gamma($1, \lambda$) law (see Section 5.3.3).

Properties

The mean and variance of the Poisson distribution with parameter λ are both equal to λ:

	Poisson(λ)
mean	λ
variance	λ
skewness	$1/\sqrt{\lambda}$
kurtosis	$3 + \lambda^{-1}$

Predictable Representation Property

We note that the PRP (5.2) in the case of the Poisson process (with intensity parameter λ) $N = \{N_t, t \geq 0\}$ can be simplified quite a lot. Since the Poisson process has only jumps of size 1, we have that in (5.2) for $i \geq 2$, $H_s^{(i)} = N_s$. In this case the infinite sum can be rewritten as a single term and we obtain that the Poisson process possesses the classical PRP. Every square integral random variable $F \in \mathcal{F}_T$ has a representation of the form,

$$F = E[F] + \int_0^T a_s \, d(N_s - \lambda s),$$

where $a = \{a_t, 0 \leq t \leq T\}$ is predictable.

5.3.2 The Compound Poisson Process

Definition

Suppose $N = \{N_t, t \geq 0\}$ is a Poisson process with intensity parameter $\lambda > 0$ and that Z_i, $i = 1, 2, \ldots$, is an i.i.d. (independent and identically distributed) sequence of random variables independent of N and following a law, L say, with characteristic function $\phi_Z(u)$. Then we say that (with the convention that an empty sum equals 0)

$$X_t = \sum_{k=1}^{N_t} Z_i, \quad t \geq 0,$$

is a compound Poisson process. The value of the process at time t, X_t, is the sum of N_t random numbers with law L. The ordinary Poisson process corresponds to the case where $Z_i = 1$, $i = 1, 2, \ldots$, i.e. where the law L is degenerate at the point 1.

Let us write (for a Borel set A) the distribution function of the law L as follows:

$$P(Z_i \in A) = \frac{\nu(A)}{\lambda},$$

where $\nu(\mathbb{R}) = \lambda < \infty$. We impose that $\nu(\{0\}) = 0$.

Then the characteristic function of X_t is given by

$$E[\exp(iu X_t)] = \exp\left(t \int_{-\infty}^{+\infty} (\exp(iux) - 1)\nu(dx) \right)$$

$$= \exp(t\lambda(\phi_Z(u) - 1)).$$

From this we can easily obtain the Lévy triplet:

$$\left[\int_{-1}^{+1} x\nu(dx), 0, \nu(dx) \right].$$

5.3.3 The Gamma Process

Definition

The density function of the Gamma distribution Gamma(a, b) with parameters $a > 0$ and $b > 0$ is given by

$$f_{\text{Gamma}}(x; a, b) = \frac{b^a}{\Gamma(a)} x^{a-1} \exp(-xb), \quad x > 0.$$

The density function clearly has a semi-heavy (right) tail. The characteristic function is given by

$$\phi_{\text{Gamma}}(u; a, b) = (1 - iu/b)^{-a}.$$

Clearly, this characteristic function is infinitely divisible. The Gamma process

$$X^{(\text{Gamma})} = \{X_t^{(\text{Gamma})}, t \geqslant 0\}$$

with parameters $a, b > 0$ is defined as the stochastic process which starts at zero and has stationary and independent Gamma distributed increments. More precisely, time enters in the first parameter: $X_t^{(\text{Gamma})}$ follows a Gamma(at, b) distribution.

The Lévy triplet of the Gamma process is given by

$$[a(1 - \exp(-b))/b, 0, a \exp(-bx)x^{-1} 1_{(x>0)} \, dx].$$

Properties

The following properties of the Gamma(a, b) distribution can easily be derived from the characteristic function:

	Gamma(a, b)
mean	a/b
variance	a/b^2
skewness	$2a^{-1/2}$
kurtosis	$3(1 + 2a^{-1})$

Note that we also have the following scaling property. If X is Gamma(a, b), then for $c > 0$, cX is Gamma$(a, b/c)$.

5.3.4 The Inverse Gaussian Process

Definition

Let $T^{(a,b)}$ be the first time a standard Brownian motion with drift $b > 0$, i.e. $\{W_s + bs, s \geqslant 0\}$, reaches the positive level $a > 0$. It is well known that this random time follows the so-called Inverse Gaussian, IG(a, b), law and has a characteristic function

$$\phi_{IG}(u; a, b) = \exp(-a(\sqrt{-2iu + b^2} - b)).$$

The IG distribution is infinitely divisible and we define the IG process $X^{(IG)} = \{X_t^{(IG)}, t \geqslant 0\}$, with parameters $a, b > 0$, as the process which starts at zero and has independent and stationary increments such that

$$E[\exp(iuX_t^{(IG)})] = \phi_{IG}(u; at, b)$$
$$= \exp(-at(\sqrt{-2iu + b^2} - b)).$$

The density function of the IG(a, b) law is explicitly known:

$$f_{IG}(x; a, b) = \frac{a}{\sqrt{2\pi}} \exp(ab)x^{-3/2} \exp(-\tfrac{1}{2}(a^2x^{-1} + b^2x)), \quad x > 0.$$

The Lévy measure of the IG(a, b) law is given by

$$\nu_{IG}(dx) = (2\pi)^{-1/2} a x^{-3/2} \exp(-\tfrac{1}{2}b^2 x) 1_{(x>0)} \, dx,$$

and the first component of the Lévy triplet equals

$$\gamma = \frac{a}{b}(2N(b) - 1),$$

where the $N(x)$ is the Normal distribution function as in (3.1).

Properties

The density is unimodal with a mode at $(\sqrt{4a^2b^2 + 9} - 3)/(2b^2)$. All positive and negative moments exist. If X follows an IG(a, b) law, we have that

$$E[X^{-\alpha}] = \left(\frac{b}{a}\right)^{2\alpha+1} E[X^{\alpha+1}], \quad \alpha \in \mathbb{R}.$$

The following characteristics can easily be obtained:

		IG(a, b)
mean		a/b
variance		a/b^3
skewness		$3(ab)^{-1/2}$
kurtosis		$3(1 + 5(ab)^{-1})$

The IG distribution satisfies the following scaling property. If X is IG(a, b), then, for a positive c, cX is IG$(\sqrt{c}a, b/\sqrt{c})$.

Origin

The name 'Inverse Gaussian' was first applied to a certain class of distributions by Tweedie (1947), who noted the inverse relationship between the cumulant generating functions of these distributions and those of Gaussian distributions. The same class of distributions was derived by Wald (1947).

5.3.5   The Generalized Inverse Gaussian Process

Definition

The Inverse Gaussian IG(a, b) law can be generalized to what is called the Generalized Inverse Gaussian distribution GIG(λ, a, b). This distribution on the positive half-line is given in terms of its density function:

$$f_{\text{GIG}}(x; \lambda, a, b) = \frac{(b/a)^\lambda}{2K_\lambda(ab)} x^{\lambda-1} \exp(-\tfrac{1}{2}(a^2 x^{-1} + b^2 x)), \quad x > 0.$$

The parameters λ, a and b are such that $\lambda \in \mathbb{R}$ while a and b are both nonnegative and not simultaneously 0.

The characteristic function is given by

$$\phi_{\text{GIG}}(u; \lambda, a, b) = \frac{1}{K_\lambda(ab)}(1 - 2iu/b^2)^{\lambda/2} K_\lambda(ab\sqrt{1 - 2iub^{-2}}),$$

where $K_\lambda(x)$ denotes the modified Bessel function of the third kind with index λ (see Appendix A).

It was shown by Barndorff-Nielsen and Halgreen (1977) that the distribution is infinitely divisible. We can thus define the GIG process as the Lévy process where the increment over the interval $[s, s + t]$, $s, t \geqslant 0$, has characteristic function $(\phi_{\text{GIG}}(u; \lambda, a, b))^t$.

The Lévy measure is rather involved and has a density on the positive real line given by

$$u(x) = x^{-1}\exp(-\tfrac{1}{2}b^2 x)\left(a^2 \int_0^{\infty} \exp(-xz)g(z)\,dz + \max\{0, \lambda\}\right),$$

where

$$g(z) = (\pi^2 a^2 z(J^2_{|\lambda|}(a\sqrt{2z}) + N^2_{|\lambda|}(a\sqrt{2z})))^{-1}$$

and where J_ν and N_ν are Bessel functions (see Appendix A). This formula for the Lévy density can, for example, be found in Barndorff-Nielsen and Shephard (2001a).

Properties

The moments of a random variable X following a GIG(λ, a, b) distribution are given by

$$E[X^k] = \left(\frac{a}{b}\right)^k \frac{K_{\lambda+k}(ab)}{K_\lambda(ab)}, \quad k \in \mathbb{R}.$$

From this it easily follows that

GIG(λ, a, b)	
mean	$aK_{\lambda+1}(ab)/(bK_\lambda(ab))$
variance	$a^2 b^{-2}K_\lambda^{-2}(ab)(K_{\lambda+2}(ab)K_\lambda(ab) + K_{\lambda+1}^2(ab))$

Special Cases

The IG(a, b) distribution. For $\lambda = -1/2$ the GIG(λ, a, b) reduces to the IG(a, b) distribution. This can easily be seen by noting that

$$K_{-1/2}(x) = \sqrt{\pi/2}\, x^{-1/2}\exp(-x).$$

The Gamma$(\tilde{a}, \tilde{b})$ distribution. For $a = 0$, $\lambda = \tilde{a} > 0$ and $b = \sqrt{2\tilde{b}}$, we obtain the Gamma$(\tilde{a}, \tilde{b})$ distribution.

Origin

The GIG was covered by Good (1953) and has been used by Sichel (1974, 1975) to construct mixtures of Poisson distributions. The GIG distribution was also used by Wise (1975) and Marcus (1975). Barndorff-Nielsen (1977, 1978) obtained the GH distribution (see below) as a mixture of the Normal distribution and the GIG distribution. Blæsild (1978) has computed moments and cumulants. Halgreen (1979) proved that the distribution is self-decomposable. A standard reference work for the GIG distribution is Jørgensen (1982).

5.3.6   The Tempered Stable Process

Definition

The characteristic function of the Tempered Stable distribution law, TS(κ, a, b), $a > 0$, $b \geqslant 0$ and $0 < \kappa < 1$, is given by

$$\phi_{TS}(u; \kappa, a, b) = \exp(ab - a(b^{1/\kappa} - 2iu)^{\kappa}).$$

This distribution is infinitely divisible and we can define the TS process

$$X^{(TS)} = \{X_t^{(TS)}, t \geqslant 0\}$$

as the process which starts at zero, has independent and stationary increments and for which the increment $X_{s+t}^{(TS)} - X_s^{(TS)}$ follows a TS(κ, ta, b) law over the time interval $[s, t+s]$.

From the characteristic function we can derive the Lévy measure of the TS process:

$$\nu_{TS}(dx) = a2^{\kappa} \frac{\kappa}{\Gamma(1-\kappa)} x^{-\kappa-1} \exp(-\tfrac{1}{2}b^{1/\kappa}x) 1_{\{x>0\}} \, dx.$$

The process is a subordinator and has infinite activity. The first term of the Lévy triplet is given by

$$\gamma = a2^{\kappa} \frac{\kappa}{\Gamma(1-\kappa)} \int_0^1 x^{-\kappa} \exp(-\tfrac{1}{2}b^{1/\kappa}x) \, dx.$$

Properties

The density is not generally known. Only the following series representation is known in general:

$$f_{TS}(x; \kappa, a, b) = \exp(ab)\exp(-\tfrac{1}{2}b^{1/\kappa}x)$$

$$\times \frac{1}{2\pi a^{1/\kappa}} \sum_{k=1}^{\infty} (-1)^{k-1} \sin(k\pi\kappa) \frac{\Gamma(k\kappa+1)}{k!} 2^{k\kappa+1} \left(\frac{x}{a^{1/\kappa}}\right)^{-k\kappa-1}.$$

The following characteristics can easily be obtained:

	TS(κ, a, b)
mean	$2a\kappa b^{(\kappa-1)/\kappa}$
variance	$4a\kappa(1-\kappa)b^{(\kappa-2)/\kappa}$
skewness	$(\kappa-2)(ab\kappa(1-\kappa))^{-1/2}$
kurtosis	$3 + (4\kappa - 6 - \kappa(1-\kappa))(ab\kappa(1-\kappa))^{-1}$

Special Cases

The IG(a, b) distribution.   For $\kappa = 1/2$ the TS(κ, a, b) reduces to the IG(a, b) distribution.

The Gamma(*a*, *b*) distribution. For the limiting case $\kappa \to 0$, we obtain the Gamma(*a*, *b*) distribution.

Origin

The class of the TS distributions was proposed by Tweedie (1984). Hougaard (1986) discussed their use in survival analysis. In Barndorff-Nielsen and Shephard (2003a) the class TS distributions were generalized to the so-called class of Modified Stable (MS) distributions of which the GIG distributions also form a subclass. They surmise that all MS distributions are infinitely divisible and in fact self-decomposable. However, a general proof is not available yet.

5.3.7 The Variance Gamma Process

Definition

The characteristic function of the VG(σ, ν, θ) law is given by

$$\phi_{VG}(u; \sigma, \nu, \theta) = (1 - iu\theta\nu + \tfrac{1}{2}\sigma^2\nu u^2)^{-1/\nu}.$$

This distribution is infinitely divisible and we can define the VG process $X^{(VG)} = \{X_t^{(VG)}, t \geq 0\}$ as the process which starts at zero, has independent and stationary increments and for which the increment $X_{s+t}^{(VG)} - X_s^{(VG)}$ follows a VG($\sigma\sqrt{t}$, ν/t, $t\theta$) law over the time interval $[s, t+s]$. Clearly (take $s = 0$ and note that $X_0^{(VG)} = 0$),

$$E[\exp(iuX_t^{(VG)})] = \phi_{VG}(u; \sigma\sqrt{t}, \nu/t, t\theta)$$
$$= (\phi_{VG}(u; \sigma, \nu, \theta))^t$$
$$= (1 - iu\theta\nu + \tfrac{1}{2}\sigma^2\nu u^2)^{-t/\nu}.$$

Madan *et al.* (1998) showed that the VG process may also be expressed as the difference of two independent Gamma processes.

This characterization allows the Lévy measure to be determined:

$$\nu_{VG}(dx) = \begin{cases} C \exp(Gx)|x|^{-1}\,dx, & x < 0, \\ C \exp(-Mx)x^{-1}\,dx, & x > 0, \end{cases}$$

where

$$C = 1/\nu > 0,$$
$$G = (\sqrt{\tfrac{1}{4}\theta^2\nu^2 + \tfrac{1}{2}\sigma^2\nu} - \tfrac{1}{2}\theta\nu)^{-1} > 0,$$
$$M = (\sqrt{\tfrac{1}{4}\theta^2\nu^2 + \tfrac{1}{2}\sigma^2\nu} + \tfrac{1}{2}\theta\nu)^{-1} > 0.$$

With this parametrization, it is clear that

$$X_t^{(VG)} = G_t^{(1)} - G_t^{(2)},$$

where $G^{(1)} = \{G_t^{(1)}, t \geqslant 0\}$ is a Gamma processes with parameters $a = C$ and $b = M$, whereas $G^{(2)} = \{G_t^{(2)}, t \geqslant 0\}$ is an independent Gamma process with parameters $a = C$ and $b = G$.

The Lévy measure has infinite mass, and hence a VG process has infinitely many jumps in any finite time interval. Since

$$\int_{-1}^{1} |x| \nu_{\text{VG}}(dx) < \infty,$$

a VG process has paths of finite variation. A VG process has no Brownian component and its Lévy triplet is given by $[\gamma, 0, \nu_{\text{VG}}(dx)]$, where

$$\gamma = \frac{-C(G(\exp(-M) - 1) - M(\exp(-G) - 1))}{MG}.$$

With the parametrization in terms of C, G and M, the characteristic function of $X_1^{(\text{VG})}$ reads as follows:

$$\phi_{\text{VG}}(u; C, G, M) = \left(\frac{GM}{GM + (M - G)iu + u^2} \right)^C.$$

In this notation we will refer to the distribution by VG(C, G, M).

Properties

Another way of defining a VG process is by seeing it as (Gamma) time-changed Brownian motion with drift. More precisely, let $G = \{G_t, t \geqslant 0\}$ be a Gamma process with parameters $a = 1/\nu > 0$ and $b = 1/\nu > 0$. Let $W = \{W_t, t \geqslant 0\}$ denote a standard Brownian motion, let $\sigma > 0$ and $\theta \in \mathbb{R}$; then the VG process $X^{(\text{VG})} = \{X_t^{(\text{VG})}, t \geqslant 0\}$, with parameters $\sigma > 0$, $\nu > 0$ and θ, can alternatively be defined as

$$X_t^{(\text{VG})} = \theta G_t + \sigma W_{G_t}.$$

When $\theta = 0$, then $G = M$ and the distribution is symmetric. Negative values of θ lead to the case where $G > M$, resulting in negative skewness. Similarly, the parameter $\nu = 1/C$ primarily controls the kurtosis:

	VG(σ, ν, θ)	VG($\sigma, \nu, 0$)
mean	θ	0
variance	$\sigma^2 + \nu\theta^2$	σ^2
skewness	$\theta\nu(3\sigma^2 + 2\nu\theta^2)/(\sigma^2 + \nu\theta^2)^{3/2}$	0
kurtosis	$3(1 + 2\nu - \nu\sigma^4(\sigma^2 + \nu\theta^2)^{-2})$	$3(1 + \nu)$

In terms of the CGM parameters this reads as follows:

	VG(C, G, M)	VG(C, G, G)
mean	$C(G - M)/(MG)$	0
variance	$C(G^2 + M^2)/(MG)^2$	$2CG^{-2}$
skewness	$2C^{-1/2}(G^3 - M^3)/(G^2 + M^2)^{3/2}$	0
kurtosis	$3(1 + 2C^{-1}(G^4 + M^4)/(M^2 + G^2)^2)$	$3(1 + C^{-1})$

Origin

The class of VG distributions was introduced by Madan and Seneta (1987) in the late 1980s as a model for stock returns. They considered (along with Madan and Seneta (1990) and Madan and Milne (1991)) the symmetric case ($\theta = 0$). Madan *et al.* (1998) treated the general case with skewness.

5.3.8   The Normal Inverse Gaussian Process

Definition

The Normal Inverse Gaussian (NIG) distribution with parameters $\alpha > 0$, $-\alpha < \beta < \alpha$ and $\delta > 0$, NIG(α, β, δ), has a characteristic function (see Barndorff-Nielsen 1995) given by

$$\phi_{\text{NIG}}(u; \alpha, \beta, \delta) = \exp(-\delta(\sqrt{\alpha^2 - (\beta + iu)^2} - \sqrt{\alpha^2 - \beta^2})).$$

Again, we can clearly see that this is an infinitely divisible characteristic function. Hence we can define the NIG process

$$X^{(\text{NIG})} = \{X_t^{(\text{NIG})}, t \geqslant 0\}$$

with $X_0^{(\text{NIG})} = 0$ stationary and independent NIG distributed increments. To be precise, $X_t^{(\text{NIG})}$ has an NIG($\alpha, \beta, t\delta$) law.

The Lévy measure for the NIG process is given by

$$\nu_{\text{NIG}}(dx) = \frac{\delta\alpha}{\pi} \frac{\exp(\beta x) K_1(\alpha|x|)}{|x|} \, dx,$$

where $K_\lambda(x)$ denotes the modified Bessel function of the third kind with index λ (see Appendix A).

An NIG process has no Brownian component and its Lévy triplet is given by $[\gamma, 0, \nu_{\text{NIG}}(dx)]$, where

$$\gamma = \frac{2\delta\alpha}{\pi} \int_0^1 \sinh(\beta x) K_1(\alpha x) \, dx.$$

The density of the NIG(α, β, δ) distribution is given by

$$f_{\text{NIG}}(x; \alpha, \beta, \delta) = \frac{\alpha\delta}{\pi} \exp(\delta\sqrt{\alpha^2 - \beta^2} + \beta x) \frac{K_1(\alpha\sqrt{\delta^2 + x^2})}{\sqrt{\delta^2 + x^2}}.$$

Properties

We can relate the NIG process to an Inverse Gaussian time-changed Brownian motion. Let $W = \{W_t, t \geqslant 0\}$ be a standard Brownian motion and let $I = \{I_t, t \geqslant 0\}$ be an IG process with parameters $a = 1$ and $b = \delta\sqrt{\alpha^2 - \beta^2}$, with $\alpha > 0$, $-\alpha < \beta < \alpha$ and $\delta > 0$; then we can show that the stochastic process

$$X_t = \beta\delta^2 I_t + \delta W_{I_t}$$

is an NIG process with parameters α, β and δ.

If a random variable X is following an NIG(α, β, δ) distribution, we have that $-X$ is NIG$(\alpha, -\beta, \delta)$ distributed. If $\beta = 0$, the distribution is symmetric. This can easily be seen from the following characteristics of the NIG distribution:

	NIG(α, β, δ)	NIG$(\alpha, 0, \delta)$
mean	$\delta\beta/\sqrt{\alpha^2 - \beta^2}$	0
variance	$\alpha^2\delta(\alpha^2 - \beta^2)^{-3/2}$	δ/α
skewness	$3\beta\alpha^{-1}\delta^{-1/2}(\alpha^2 - \beta^2)^{-1/4}$	0
kurtosis	$3\left(1 + \dfrac{\alpha^2 + 4\beta^2}{\delta\alpha^2\sqrt{\alpha^2 - \beta^2}}\right)$	$3(1 + \delta^{-1}\alpha^{-1})$

The NIG distributions have semi-heavy tails, in particular

$$f_{\text{NIG}}(x; \alpha, \beta, \delta) \sim |x|^{-3/2} \exp((\mp\alpha + \beta)x) \quad \text{as } x \to \pm\infty,$$

up to a multiplicative constant.

Origin

The NIG distribution was introduced by Barndorff-Nielsen (1995). See also Barndorff-Nielsen (1997) and Rydberg (1996a,b, 1997a).

5.3.9 The CGMY Process

Definition

The CGMY(C, G, M, Y) distribution is a four-parameter distribution, with characteristic function

$$\phi_{\text{CGMY}}(u; C, G, M, Y) = \exp(C\Gamma(-Y)((M - \mathrm{i}u)^Y - M^Y + (G + \mathrm{i}u)^Y - G^Y)).$$

The CGMY distribution is infinitely divisible and we can define the CGMY Lévy process

$$X^{(\text{CGMY})} = \{X_t^{(\text{CGMY})}, t \geq 0\}$$

as the process which starts at zero, has independent and stationary distributed increments and in which the increment over a time interval of length s follows a CGMY(sC, G, M, Y) distribution; in other words, the characteristic function of $X_t^{(\text{CGMY})}$ is given by

$$
\begin{aligned}
E[\exp(iu X_t^{(\text{CGMY})})] &= \phi_{\text{CGMY}}(u; tC, G, M, Y) \\
&= (\phi_{\text{CGMY}}(u; C, G, M, Y))^t \\
&= \exp(Ct\Gamma(-Y)((M - iu)^Y - M^Y + (G + iu)^Y - G^Y)).
\end{aligned}
$$

The Lévy measure for the CGMY process is given by

$$
\nu_{\text{CGMY}}(dx) = \begin{cases} C \exp(Gx)(-x)^{-1-Y}\,dx, & x < 0, \\ C \exp(-Mx)x^{-1-Y}\,dx, & x > 0. \end{cases}
$$

The first parameter of the Lévy triplet equals

$$
\gamma = C\left(\int_0^1 \exp(-Mx)x^{-Y}\,dx - \int_{-1}^0 \exp(Gx)|x|^{-Y}\,dx \right).
$$

The range of the parameters is restricted to $C, G, M > 0$ and $Y < 2$. Choosing the Y parameters greater than or equal to two does not yield a valid Lévy measure.

Properties

The following characteristics of the CGMY distribution can easily be calculated:

	CGMY(C, G, M, Y)
mean	$C(M^{Y-1} - G^{Y-1})\Gamma(1 - Y)$
variance	$C(M^{Y-2} + G^{Y-2})\Gamma(2 - Y)$
skewness	$\dfrac{C(M^{Y-3} - G^{Y-3})\Gamma(3 - Y)}{(C(M^{Y-2} + G^{Y-2})\Gamma(2 - Y))^{3/2}}$
kurtosis	$3 + \dfrac{C(M^{Y-4} + G^{Y-4})\Gamma(4 - Y)}{(C(M^{Y-2} + G^{Y-2})\Gamma(2 - Y))^2}$

The CGMY process is a pure jump process, that is, it contains no Brownian part. The path behaviour is determined by the Y parameters. If $Y < 0$, the paths have finite jumps in any finite interval; if not, the paths have infinitely many jumps in any finite time interval, i.e. the process has infinite activity. Moreover, if the Y parameters lie in the interval $[1, 2)$, the process is of infinite variation.

Special Cases

The VG(C, G, M) distribution. The Variance Gamma distribution/process is a special case of the CGMY distribution/process. If $Y = 0$, the CGMY reduces to VG: CGMY(C, G, M, 0) = VG(C, G, M).

Origin

In order to obtain a more flexible process than the VG process—one allowing finite activity, infinite activity and infinite variation—the additional parameter Y was introduced by Carr *et al.* (2002).

In Carr *et al.* (2002) the above four-parameter case was studied and later generalized to a six-parameter case in Carr *et al.* (2003). The C and Y parameters are split into C_n and C_p and into Y_p and Y_n, corresponding to the positive ('p') and negative ('n') parts in the Lévy measure. We restrict ourselves here to the four-parameter case.

The above family of distributions is also called the KoBoL family by some authors (after Koponen (1995) and Boyarchenko and Levendorskiĭ (1999); see also Bouchaud and Potters (1997), Cont *et al.* (1997), Matacz (1997) and Boyarchenko and Levendorskiĭ (2000, 2002b)). Initially, the name (generalized) TLP (truncated Lévy process) was used. However, Shiryaev remarked it was misleading, and it was then replaced by the name KoBoL.

5.3.10 The Meixner Process

Definition

The density of the Meixner distribution (Meixner(α, β, δ)) is given by

$$f_{\text{Meixner}}(x; \alpha, \beta, \delta) = \frac{(2 \cos(\beta/2))^{2\delta}}{2\alpha\pi\Gamma(2d)} \exp\left(\frac{bx}{a}\right)\left|\Gamma\left(\delta + \frac{ix}{\alpha}\right)\right|^2,$$

where $\alpha > 0$, $-\pi < \beta < \pi$, $\delta > 0$.

The characteristic function of the Meixner(α, β, δ) distribution is given by

$$\phi_{\text{Meixner}}(u; \alpha, \beta, \delta) = \left(\frac{\cos(\beta/2)}{\cosh((\alpha u - i\beta)/2)}\right)^{2\delta}.$$

The Meixner(α, β, δ) distribution is infinitely divisible: $\phi_{\text{Meixner}}(u; \alpha, \beta, \delta) = (\phi_{\text{Meixner}}(u; \alpha, \beta, \delta/n))^n$. We can thus associate with it a Lévy process, which we call the Meixner process. More precisely, a Meixner process

$$X^{(\text{Meixner})} = \{X_t^{(\text{Meixner})}, t \geqslant 0\}$$

is a stochastic process which starts at zero, i.e. $X_0^{(\text{Meixner})} = 0$, has independent and stationary increments, and where the distribution of $X_t^{(\text{Meixner})}$ is given by the Meixner distribution Meixner(α, β, δt).

We can show (see Grigelionis 1999) that our Meixner process

$$X^{(\text{Meixner})} = \{X_t^{(\text{Meixner})}, t \geqslant 0\}$$

has no Brownian part and a pure jump part governed by the Lévy measure

$$v(\mathrm{d}x) = \delta \frac{\exp(\beta x/\alpha)}{x \sinh(\pi x/\alpha)} \, \mathrm{d}x.$$

The first parameter in the Lévy triplet equals

$$\gamma = \alpha\delta \tan(\beta/2) - 2\delta \int_1^\infty \frac{\sinh(\beta x/\alpha)}{\sinh(\pi x/\alpha)} \, \mathrm{d}x.$$

Because $\int_{-1}^{+1} |x| v(\mathrm{d}x) = \infty$, the process is of infinite variation.

Properties

Moments of all order of this distribution exist. If a random variable X is following a Meixner(α, β, δ) distribution, we have that $-X$ is Meixner$(\alpha, -\beta, \delta)$ distributed. Next, we give some relevant quantities for the general case and the symmetric case around zero, i.e. with $\beta = 0$:

	Meixner(α, β, δ)	Meixner$(\alpha, 0, \delta)$
mean	$\alpha\delta \tan(\beta/2)$	0
variance	$\frac{1}{2}\alpha^2\delta(\cos^{-2}(\beta/2))$	$\frac{1}{2}\alpha^2\delta$
skewness	$\sin(\beta/2)\sqrt{2/\delta}$	0
kurtosis	$3 + (2 - \cos(\beta))/\delta$	$3 + 1/\delta$

We can clearly see that the kurtosis of the Meixner distribution is always greater than the Normal kurtosis, which always equals 3.

The Meixner(α, β, δ) distribution has semi-heavy tails (see Grigelionis 2000),

$$f_{\text{Meixner}}(x; \alpha, \beta, \delta) \sim \begin{cases} C_-|x|^{\rho_-} \exp(-\eta_-|x|) & \text{as } x \to -\infty, \\ C_+|x|^{\rho_+} \exp(-\eta_+|x|) & \text{as } x \to +\infty, \end{cases}$$

where

$$\rho_- = \rho_+ = 2\delta - 1, \qquad \eta_- = (\pi - \beta)/\alpha, \qquad \eta_+ = (\pi + \beta)/\alpha.$$

and for some $C_-, C_+ \geqslant 0$.

The Meixner process is related to the process studied by Biane *et al.* (2001) (see also Pitman and Yor 2000),

$$\chi_t = \frac{2}{\pi^2} \sum_{n=1}^\infty \frac{\Gamma_{n,t}}{(n - 1/2)^2},$$

for a sequence of independent Gamma Processes (with $a = 1$ and $b = 1$) $\Gamma_{n,t}$, i.e. a
Lévy process with $E[\exp(i\theta\Gamma_{n,t})] = (1 - i\theta)^{-t}$.

Biane *et al.* (2001) showed that χ_t has characteristic function

$$E[\exp(iu\chi_t)] = \left(\frac{1}{\cosh\sqrt{-2ui}}\right)^t.$$

Let $W = \{W_t, t \geqslant 0\}$ be a standard Brownian motion, then the Brownian time change
W_{χ_t} has characteristic function

$$E[\exp(iu W_{\chi_t})] = \left(\frac{1}{\cosh u}\right)^t,$$

or, equivalently, W_{χ_t} follows a Meixner(2, 0, t) distribution.

Origin

The Meixner process was introduced in Schoutens and Teugels (1998) (see also
Schoutens 2000) and Grigelionis (1999) later suggested that it serve for fitting stock
returns. This application to finance was worked out in Schoutens (2001, 2002).

The Meixner process originates from the theory of orthogonal polynomials. The
Meixner(1, $2\zeta - \pi$, δ) distribution is the measure of orthogonality of the Meixner–
Pollaczek polynomials $\{P_n(x; \delta, \zeta), n = 0, 1, \ldots\}$ (for a definition see Koekoek and
Swarttouw (1998) or Appendix A). Moreover, the monic (i.e. with leading coefficient
equal to 1) Meixner–Pollaczek polynomials $\{\tilde{P}_n(x; \delta, \zeta), n = 0, 1, \ldots\}$ are martin-
gales for the Meixner process ($\alpha = 1, \delta = 1, \zeta = (\beta + \pi)/2$):

$$E[\tilde{P}_n(X_t^{(\text{Meixner})}; t, \zeta) \mid X_s^{(\text{Meixner})}] = \tilde{P}_n(X_s^{(\text{Meixner})}; s, \zeta).$$

Note the similarity with the classical martingale relation between the standard Brownian motion $\{W_t, t \geqslant 0\}$ and the Hermite polynomials $\{H_n(x; \sigma), n = 0, 1, \ldots\}$ (see
Schoutens (2000) or Appendix A for a definition):

$$E[\tilde{H}_n(W_t; t) \mid W_s] = \tilde{H}_n(W_s; s).$$

The Meixner distribution is a special case of the Generalized z (GZ) distributions,
which were later defined in Grigelionis (2000) and have a characteristic function of
the form,

$$\phi_{\text{GZ}}(u; \alpha, \beta_1, \beta_2, \delta) = \left(\frac{B(\beta_1 + i\alpha u/2\pi, \beta_2 - i\alpha u/2\pi)}{B(\beta_1, \beta_2)}\right)^{2\delta},$$

where $\alpha, \beta_1, \beta_2, \delta > 0$. For

$$\beta_1 = \frac{1}{2} + \frac{\beta}{2\pi} \quad \text{and} \quad \beta_2 = \frac{1}{2} - \frac{\beta}{2\pi},$$

we obtain the Meixner process.

5.3.11   The Generalized Hyperbolic Process

Definition

The Generalized Hyperbolic (GH) distribution GH(α, β, δ, v) is defined in Barndorff-Nielsen (1977) through its characteristic function,

$$\phi_{GH}(u; \alpha, \beta, \delta, v) = \left(\frac{\alpha^2 - \beta^2}{\alpha^2 - (\beta + iu)^2}\right)^{v/2} \frac{K_v(\delta\sqrt{\alpha^2 - (\beta + iu)^2})}{K_v(\delta\sqrt{\alpha^2 - \beta^2})},$$

where K_v is the modified Bessel function (see Appendix A).

The density of the GH(α, β, δ, v) distribution is given by

$$f_{GH}(x; \alpha, \beta, \delta, v) = a(\alpha, \beta, \delta, v)(\delta^2 + x^2)^{(v-1/2)/2} K_{v-1/2}(\alpha\sqrt{\delta^2 + x^2}) \exp(\beta x),$$

$$a(\alpha, \beta, \delta, v) = \frac{(\alpha^2 - \beta^2)^{v/2}}{\sqrt{2\pi}\alpha^{v-1/2}\delta^v K_v(\delta\sqrt{\alpha^2 - \beta^2})},$$

where

$$\begin{aligned}
\delta &\geqslant 0, \quad |\beta| < \alpha \quad \text{if } v > 0, \\
\delta &> 0, \quad |\beta| < \alpha \quad \text{if } v = 0, \\
\delta &> 0, \quad |\beta| \leqslant \alpha \quad \text{if } v < 0.
\end{aligned}$$

The GH distribution turns out to be infinitely divisible (see Barndorff-Nielsen and Halgreen 1977) and we can define a GH Lévy process $X^{(GH)} = \{X_t^{(GH)}, t \geqslant 0\}$ as the stationary process which starts at zero and has independent increments and where the distribution of $X_t^{(GH)}$ has characteristic function,

$$E[\exp(iu X_t^{(GH)})] = (\phi_{GH}(u; \alpha, \beta, \delta, v))^t.$$

The Lévy measure $v(dx)$ for the GH process is rather involved:

$$v(dx) = \begin{cases} \dfrac{\exp(\beta x)}{|x|}\left(\displaystyle\int_0^\infty \dfrac{\exp(-|x|\sqrt{2y + \alpha^2})}{\pi^2 y(J_v^2(\delta\sqrt{2y}) + N_v^2(\delta\sqrt{2y}))}\, dy + v\exp(-\alpha|x|)\right), \\
\hspace{9cm} v \geqslant 0, \\[4pt]
\dfrac{\exp(\beta x)}{|x|}\displaystyle\int_0^\infty \dfrac{\exp(-|x|\sqrt{2y + \alpha^2})}{\pi^2 y(J_{-v}^2(\delta\sqrt{2y}) + N_{-v}^2(\delta\sqrt{2y}))}\, dy, \\
\hspace{9cm} v < 0, \end{cases}$$

where the functions J_v and N_v are the Bessel functions defined in Appendix A.

Properties

The GH distributions have semi-heavy tails, in particular,

$$f_{GH}(x; \alpha, \beta, \delta, v) \sim |x|^{v-1} \exp((\mp\alpha + \beta)x) \quad \text{as } x \to \pm\infty,$$

up to a multiplicative constant.

The generalized hyperbolic distribution has the following mean and variance:

GH$(\alpha, \beta, \delta, v)$	
mean	$\beta\delta(\alpha^2 - \beta^2)^{-1}K_{v+1}(\zeta)K_v^{-1}(\zeta)$
variance	$\delta^2\left(\dfrac{K_{v+1}(\zeta)}{\zeta K_v(\zeta)} + \dfrac{\beta^2}{\alpha^2 - \beta^2}\left(\dfrac{K_{v+2}(\zeta)}{K_v(\zeta)} - \dfrac{K_{v+1}^2(\zeta)}{K_v^2(\zeta)}\right)\right)$

where $\zeta = \delta\sqrt{\alpha^2 - \beta^2}$.

The GH distribution can also be represented as a Normal variance-mean mixture:

$$f_{GH}(x; \alpha, \beta, \delta, v) = \int_0^\infty f_{Normal}(x; \mu + \beta w, w) f_{GIG}(w; v, \delta, \sqrt{\alpha^2 - \beta^2})\, dw.$$

Special Cases

Some of the above processes are special cases of the GH process.

The Variance Gamma Process. This process can be obtained from the GH processes by taking $v = \sigma^2/v$, $\alpha = \sqrt{(2/v) + (\theta^2/\sigma^4)}$, $\beta = \theta/\sigma^2$ and $\delta \to 0$.

The Hyperbolic Process. For $v = 1$, we obtain the Hyperbolic process (HYP), where $X_1^{(HYP)}$ is following the Hyberbolic distribution HYP(α, β, δ) with characteristic function:

$$\phi_{HYP}(u; \alpha, \beta, \delta) = \left(\frac{\alpha^2 - \beta^2}{\alpha^2 - (\beta + iu)^2}\right)^{1/2} \frac{K_1(\delta\sqrt{\alpha^2 - (\beta + iu)^2})}{K_1(\delta\sqrt{\alpha^2 - \beta^2})}.$$

The density reduces to

$$f_{HYP}(x; \alpha, \beta, \delta) = \frac{\sqrt{\alpha^2 - \beta^2}}{2\delta\alpha K_1(\delta\sqrt{\alpha^2 - \beta^2})} \exp(-\alpha\sqrt{\delta^2 + x^2} + \beta x).$$

The Normal Inverse Gaussian Process. For $v = -1/2$ we obtain the Normal Inverse Gaussian process: we have GH$(\alpha, \beta, \delta, -1/2) = $ NIG(α, β, δ).

Origin

The GH distributions were introduced by Barndorff-Nielsen (1977) as a model for the grain-size distribution of wind-blown sand. Two subclasses of the GH distribution were first used to model financial data. Eberlein and Keller (1995) used the Hyperbolic distribution and in the same year Barndorff-Nielsen (1995) proposed the NIG distribution. Eberlein and Prause (1998) and Prause (1999) finally studied the whole family of GH distributions as a model to describe the log returns of some financial asset. For an overview of the GH distribution and its limiting cases, see Eberlein and v. Hammerstein (2002). The Hyperbolic Model is also investigated in Bingham and Kiesel (2001a,b). For semi-parametric generalizations, see Bingham and Kiesel (2002).

5.4 Adding an Additional Drift Term

In the above VG, NIG, CGMY, Meixner and GH cases, an additional 'drift' or location parameter $m \in \mathbb{R}$ can be introduced. This parameter will play some special role in the risk-neutral modelling of our risky asset. Essentially, the transformation is completely of the same manner as the one which transforms a Normal$(0, \sigma^2)$ random variable into a Normal(m, σ^2) random variable. Moreover, this extension does not influence the infinite divisibility property nor the self-decomposability of the distribution. Only in this section we will denote the original process by $\bar{X}$ and the newly obtained one by X. The same notation will be used for the characteristic function and the ingredients of the Lévy triplet.

The (extended) distribution in the Meixner case is denoted by Meixner$(\alpha, \beta, \delta, m)$. For other distributions the new parameter will give rise to distributions we denote by VG(σ, ν, θ, m) (or VG(C, G, M, m)), NIG$(\alpha, \beta, \delta, m)$, CGMY$(C, G, M, Y, m)$, GH$(\alpha, \beta, \delta, \nu, m)$.

The new distribution has a characteristic function ϕ in terms of the original characteristic function $\bar{\phi}$:

$$\phi(u) = \bar{\phi}(u) \exp(ium).$$

This new parameter is just a translation by the value $m \in \mathbb{R}$ of the distribution. In terms of the process this means a term mt is added to the process $\bar{X}$, i.e.

$$X_t = \bar{X}_t + mt.$$

This is reflected only in the first parameter of the Lévy triplet, which now equals

$$\gamma = \bar{\gamma} + m, \qquad \sigma^2 = \bar{\sigma}^2, \qquad \nu(dx) = \bar{\nu}(dx).$$

In terms of density functions this comes down to

$$f(x) = \bar{f}(x - m).$$

In Figure 5.1 we can clearly see that the density of the Meixner distribution ($\alpha = 1$, $\beta = 0$, $\delta = 1$) is just shifted with a value m.

5.5 Examples of OU Processes

Next, we list some examples of OU processes $y = \{y_t, t \geq 0\}$. The BDLP is denoted by $z = \{z_t, t \geq 0\}$. Recall that $k(u) = k_z(u) = \log E[\exp(-uz_1)]$ is the cumulant function of z_1. For our financial models, we will only make use of the Gamma–OU processes and the IG–OU processes but here we also list other examples for completeness.

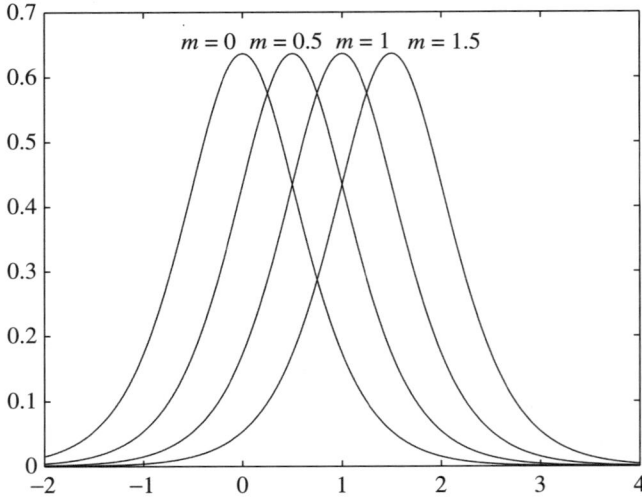

Figure 5.1   The role of the m parameter.

5.5.1   The Gamma–OU Process

The Gamma(a, b) process has a Lévy density (living on the positive real line) given by $u(x) = a \exp(-bx)/x 1_{(x>0)}$. Using the criterion (5.3), the Gamma(a, b) distribution is clearly self-decomposable. Using relation (5.5), the corresponding BDLP z has Lévy density

$$w(x) = ab \exp(-bx) 1_{(x>0)}.$$

The associated cumulant function is given by

$$k(u) = -au(b + u)^{-1}.$$

From this we can easily derive that the BDLP for the Gamma(a, b)–OU process is a compound Poisson process, i.e.

$$z_t = \sum_{n=1}^{N_t} x_n,$$

where $N = \{N_t, t \geqslant 0\}$ is a Poisson process with intensity parameter a, i.e. $E[N_t] = at$ and $\{x_n, n = 1, 2, \dots\}$ is an independent and identically distributed sequence; each x_n follows a Gamma$(1, b)$ law. Since the BDLP is compound Poisson, it only jumps a finite number of times in every compact interval. Hence, the Gamma–OU process also jumps a finite number of times in every compact (time) interval.

In the Gamma–OU case the characteristic function of the intOU process $Y_t = \int_0^t y_s\, ds$ can be given explicitly:

$$
\begin{aligned}
\phi_{\text{Gamma–OU}}&(u, t; \lambda, a, b, y_0) \\
&= E[\exp(iuY_t) \mid y_0] \\
&= \exp(iuy_0\lambda^{-1}(1 - \exp(-\lambda t))) \\
&\quad \times \exp\left(\frac{\lambda a}{iu - \lambda b}\left(b \log\left(\frac{b}{b - iu\lambda^{-1}(1 - \exp(-\lambda t))}\right) - iut\right)\right).
\end{aligned}
$$

The inverse tail mass function can also be made explicit in this case. Since $W^+(x) = a \exp(-bx)$, its inverse can be analytically expressed as

$$
W^{-1}(x) = \max\{0, -b^{-1}\log(x/a)\}.
$$

5.5.2 The IG–OU Process

Similarly as in the Gamma case, it follows from the Lévy density of the $IG(a, b)$ process,

$$
u(x) = (2\pi)^{-1/2}ax^{-3/2}\exp(-\tfrac{1}{2}b^2 x)1_{(x>0)},
$$

that the $IG(a, b)$ distribution is self-decomposable and that the Lévy density of the corresponding BDLP is

$$
w(x) = (2\pi)^{-1/2}\tfrac{1}{2}a(x^{-1} + b^2)x^{-1/2}\exp(-\tfrac{1}{2}b^2 x)1_{(x>0)}.
$$

The corresponding cumulant function is given by

$$
k(u) = -uab^{-1}(1 + 2ub^{-2})^{-1/2}.
$$

From the above expressions, it can be derived (see, for example, Barndorff-Nielsen 1998) that in the case of the $IG(a, b)$–OU process the BDLP is a sum of two independent Lévy processes $z = z^{(1)} + z^{(2)} = \{z_t = z_t^{(1)} + z_t^{(2)}, t \geq 0\}$, where $z^{(1)}$ is an IG–Lévy process with parameters $a/2$ and b, while $z^{(2)}$ is of the form,

$$
z_t^{(2)} = b^{-2}\sum_{n=1}^{N_t} v_n^2,
$$

where $N = \{N_t, t \geq 0\}$ is a Poisson process with intensity parameter $ab/2$, i.e. $E[N_t] = abt/2$, and $\{v_n, n = 1, 2, \ldots\}$ is an independent and identically distributed sequence; each v_n follows a $\text{Normal}(0, 1)$ law independent from the Poisson process N. Since the BDLP (via $z^{(1)}$) jumps infinitely often in every finite (time) interval, the IG–OU process also jumps infinitely often in every interval.

In the IG–OU case the characteristic function of the intOU process $Y_t = \int_0^t y_s \, ds$ can be given explicitly. The following expression was independently derived by Nicolato and Venardos (2003) and Tompkins and Hubalek (2000),

$$\phi_{\text{IG–OU}}(u, t; \lambda, a, b, y_0) = E[\exp(iuY_t) \mid y_0]$$

$$= \exp\left(\frac{iuy_0}{\lambda}(1 - \exp(-\lambda t)) + \frac{2aiu}{b\lambda} A(u, t)\right),$$

where

$$A(u, t) = \frac{1 - \sqrt{1 + \kappa(1 - \exp(-\lambda t))}}{\kappa}$$

$$+ \frac{1}{\sqrt{1+\kappa}}\left(\operatorname{artanh}\left(\frac{\sqrt{1 + \kappa(1 - \exp(-\lambda t))}}{\sqrt{1+\kappa}}\right) - \operatorname{artanh}\left(\frac{1}{\sqrt{1+\kappa}}\right)\right),$$

$$\kappa = -2b^{-2}iu/\lambda.$$

5.5.3 Other Examples

Here we list a series of other distributions which are self-decomposable and as such for which there exists an OU process driven by a BDLP (not necessary a subordinator).

The TS–OU Process

It follows from the Lévy density of the $TS(\kappa, a, b)$ process,

$$u(x) = a2^{\kappa} \frac{\kappa}{\Gamma(1 - \kappa)} x^{-\kappa-1} \exp(-\tfrac{1}{2}b^{1/\kappa}x)1_{(x>0)},$$

that the $TS(\kappa, a, b)$ distribution is self-decomposable. The Lévy density and cumulant function of the BDLP are given by, respectively,

$$w(x) = a2^{\kappa} \frac{\kappa}{\Gamma(1 - \kappa)}(\kappa x^{-1} + \tfrac{1}{2}b^{1/\kappa})x^{-\kappa} \exp(-\tfrac{1}{2}b^{1/\kappa}x)1_{(x>0)},$$

$$k(u) = -2ua^{2\kappa}\kappa b^{-1}(b^2 + 2u)^{\kappa-1}.$$

This shows (see Barndorff-Nielsen and Shephard 2003a) that the BDLP of the TS–OU process is the sum of a $TS(\kappa, \kappa a, b)$ Lévy process $z^{(1)}$ plus a compound Poisson process $z^{(2)}$, with Lévy density

$$a2^{\kappa-1} \frac{\kappa}{\Gamma(1 - \kappa)} b^{1/\kappa} x^{-\kappa} \exp(-\tfrac{1}{2}b^{1/\kappa}x)1_{(x>0)},$$

or, in other words,

$$z_t^{(2)} = \sum_{k=1}^{N_t} x_i, \quad t \geq 0,$$

where N_t is a Poisson process with intensity parameter $ab\kappa$ and x_i are independent $\text{Gamma}(1 - \kappa, b^{1/\kappa}/2)$ random variables.

The Meixner–OU Process

The Meixner$(\alpha, \beta, \delta, m)$ is self-decomposable (see Grigelionis 1999). We have

$$w(x) = \delta\lambda((\pi - \beta)\exp((\beta + \pi)x/\alpha)$$
$$+ (\pi + \beta)\exp((\beta - \pi)x/\alpha))(\sinh(\pi x/\alpha))^{-2},$$
$$k(u) = \alpha\delta\lambda u\tan((\alpha u - \beta)/2) - \lambda m.$$

The Meixner–OU process is *not* driven by a BDLP that is a subordinator. The BDLP has a Lévy density that lives over the whole real line. This means that the Meixner–OU process (and its BDLP) can jump upwards and downwards.

The NIG–OU Process

The NIG(α, β, δ) is self-decomposable (see Barndorff-Nielsen 1998). We have

$$w(x) = \pi^{-1}\delta\alpha[(|x|^{-1} - \beta\,\mathrm{sgn}(x))\mathrm{K}_1(\alpha|x|) + \alpha\mathrm{K}_0(\alpha|x|)]\exp(\beta x).$$

Thus, the NIG–OU process is also *not* driven by a BDLP that is a subordinator. The BDLP has a Lévy density that lives over the whole real line and the NIG–OU process (and its BDLP) can jump upwards and downwards.

Since, if a random variable X is following an NIG(α, β, δ) distribution, we have that $-X$ is NIG$(\alpha, -\beta, \delta)$ distributed; we assume here $\beta > 0$.

In this case Barndorff-Nielsen (1998) proved that the BDLP z is the sum of three independent Lévy processes:

$$z = z^{(1)} + z^{(2)} + z^{(3)} = \{z_t = z_t^{(1)} + z_t^{(2)} + z_t^{(3)}, t \geqslant 0\}.$$

Let $\rho = \beta/\alpha$.

The first process $z^{(1)}$ is the NIG Lévy processes with an NIG$(\alpha, \beta, (1 - \rho)\delta)$ law at time 1, the second process $z^{(2)}$ has the form,

$$z^{(2)} = \frac{1}{\alpha\sqrt{1 - \rho^2}}\sum_{n=1}^{N_t}(v_n^2 - \tilde{v}_n^2),$$

where $N = \{N_t, t \geqslant 0\}$ denotes a Poisson process with intensity parameter

$$\frac{1}{\delta\alpha\sqrt{(1 - \rho)/(1 + \rho)}}$$

and $\{v_n, n = 1, 2, \ldots\}$ and $\{\tilde{v}_n, n = 1, 2, \ldots\}$ are independent standard Normally, i.e. Normal$(0, 1)$, distributed sequences independent of the Poisson process N. Finally, the moment-generating function $\vartheta(u) = E[\exp(uz_t^{(3)})]$ of the third process $z^{(3)}$ is given by

$$\vartheta(u) = \exp(t\rho\delta(\beta\sqrt{(\alpha - \beta)/(\alpha + \beta)} - (u + \beta)\sqrt{(\alpha - u - \beta)/(\alpha + u + \beta)})).$$

6

Stock Price Models Driven by Lévy Processes

In this chapter we will try to model stock-price behaviour by a more sophisticated stochastic process than the Brownian motion of the Black–Scholes model. The stock-price dynamics are now driven by a Lévy process. The stock-price behaviour is now modelled as the exponential of a Lévy process. We are able to take into account skewness and excess kurtosis and show that we can fit very accurately our underlying distributions to historical data. Next, we will price European options under this model. Unfortunately, as in the most realistic models, there is no unique equivalent martingale measure: the proposed Lévy models are incomplete. We thus need to choose an equivalent martingale measure to price our options. We look at two different possibilities: the Esscher transform martingale measure and a mean-correcting martingale measure. Finally, we will try to calibrate our model to a set of option prices available in the market. We clearly observe a significant improvement with respect to the Black–Scholes model.

6.1 Statistical Testing

We fit the Meixner distribution to several datasets, which we have already encountered in Chapter 4, of daily log returns of popular indices. By this we illustrate that the more flexible distributions, such as the Meixner, the VG, the NIG, the CGMY and the GH, are more suitable than the Normal distribution. Similar fits can, for example, be found for the Hyperbolic distribution in Eberlein and Keller (1995).

6.1.1 Parameter Estimation

Here we focus on how to estimate the parameters of a density function. Denote the density function by $f(x; \theta)$; θ is the set of unknown parameters to be estimated.

Lévy Processes in Finance   W. Schoutens
© 2003 John Wiley & Sons, Ltd   ISBN: 0-470-85156-2

We assume that we have n independent observations $x_1, \ldots, x_n$ of a random variable X. Typically, these observations will be the log returns of our financial asset. From these observations we would like to deduce reasonable estimators for the parameter set θ. Note that under a Lévy process setting, the log returns over nonoverlapping intervals of fixed length (typically, one day) will be independent and identically distributed. Sometimes, ad hoc methods can also deliver reasonable estimators. However, we give an overview of the classical maximum-likelihood estimation method.

Maximum-Likelihood Estimators

The maximum-likelihood estimator (MLE) $\hat{\theta}_{\text{MLE}}$ is the parameter set that maximizes the likelihood function

$$L(\theta) = \prod_{i=1}^{n} f(x_i; \theta).$$

Thus, we choose values for the parameters that maximize the chance (or likelihood) of the data occurring.

Maximizing an expression is equivalent to maximizing the logarithm of the expression, and this is sometimes easier. So, we sometimes maximize instead the log likelihood function,

$$\log L(\theta) = \sum_{i=1}^{n} \log f(x_i; \theta).$$

To maximize the (log)-likelihood function, we often have to rely on numerical procedures; however, in a few cases these estimators can be calculated explicitly. The MLE estimators for the mean and variance of the Normal distribution are given by the sample mean and sample variance:

$$\hat{\mu}_{\text{MLE}} = \frac{1}{n} \sum_{i=1}^{n} x_i, \qquad \hat{\sigma}^2_{\text{MLE}} = \frac{1}{n} \sum_{i=1}^{n} x_i^2 - \left(\frac{1}{n} \sum_{i=1}^{n} x_i \right)^2.$$

6.1.2   Statistical Testing

Density and Log Density Fits

Figure 6.1 shows the Gaussian kernel density estimator based on the daily log returns of the S&P 500 Index over the period from 1970 until the end of 2001, together with the fitted Meixner distribution, with parameters from Table 6.1. Compared with Figure 4.1, in which the Normal counterpart was plotted, we see a significant improvement. Note also that the Meixner distribution has semi-heavy tails and as such is capable of also fitting the tail behaviour quite well. This can be seen from the log density plot in Figure 6.1.

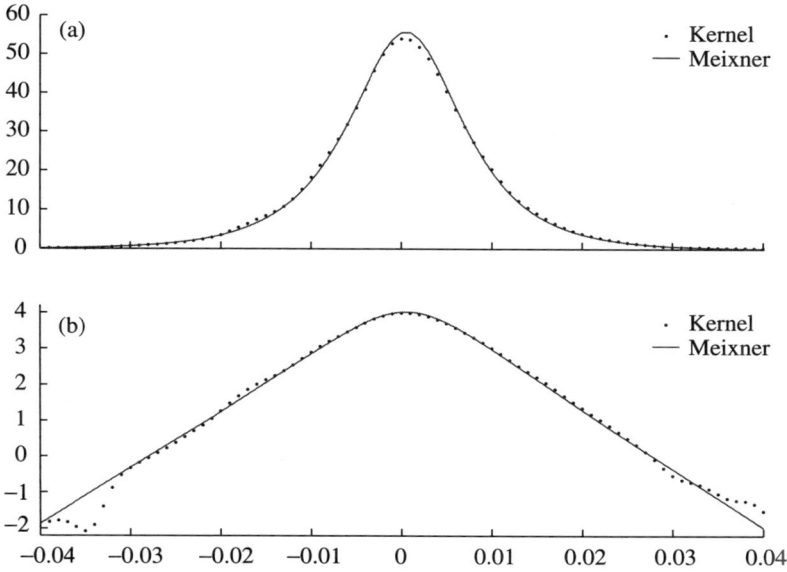

Figure 6.1 (a) Meixner density and Gaussian kernel density estimators and (b) log densities of the daily log returns of the S&P 500 Index.

Table 6.1 Meixner χ^2-test: MLE parameters and P-values.

Index	a	b	d	m	P_{Meixner}-value
S&P 500 (1970–2001)	0.0204	−0.0829	0.4140	0.0006	0.4754
S&P 500 (1997–1999)	0.0198	−0.1166	0.6923	0.0017	0.4880
DAX	0.0331	−0.0967	0.4617	0.0023	0.4937
Nasdaq-Composite	0.0301	−0.0732	0.5414	0.0020	0.1932
CAC-40	0.0252	−0.1936	0.6316	0.0029	0.2473
SMI	0.0307	−0.1231	0.4306	0.0021	0.4431

χ^2-Tests

For the χ^2-tests we take the same intervals as when we were testing the Normal distribution in Chapter 4. Parameters are estimated by the MLE method. In the Meixner case, four parameters have to be estimated, so we take $n - 5$ degrees of freedom (n is the number of observations).

Table 6.1 shows the values of the P-values of the χ^2-test statistic with equal width for the Meixner null hypothesis. Recall that we reject the hypothesis if the P-value is less than our level of significance, which we take to be 0.05, and accept it otherwise.

We see that the Meixner hypothesis is accepted and yields a very high P-value.

6.2   The Lévy Market Model

Instead of modelling the log returns with a Normal distribution, we now replace it with a more sophisticated infinitely divisible distribution. So, let $X = \{X_t, t \geq 0\}$ be a Lévy process. We assume that our market consists of one riskless asset (the bond), with a price process given by $B_t = \exp(rt)$, and one risky asset (the stock or index). The model for the risky asset is

$$S_t = S_0 \exp(X_t).$$

The log returns $\log(S_{t+s}/S_t)$ of such a model follow the distribution of increments of length s of the Lévy process X.

In the literature several particular choices for the Lévy processes have been studied in detail. Madan and Seneta (1987, 1990) have proposed a VG Lévy process. Eberlein and Keller (1995) proposed the Hyperbolic Model, and Barndorff-Nielsen (1995) the NIG model. These three models were brought together as special cases of the Generalized Hyperbolic Model, which was developed by Eberlein and co-workers in a series of papers (Eberlein and Prause 1998; Eberlein et al. 1998; Prause 1999). Carr et al. (2002) introduced the CGMY model; this family of distributions is also called the KoBoL family by some authors, referring to Koponen (1995) and Boyarchenko and Levendorskiĭ (1999) (see also Bouchaud and Potters 1997; Boyarchenko and Levendorskiĭ 2000, 2002b; Cont et al. 1997; Matacz 1997). Finally, the Meixner model was used in Schoutens (2001).

An accessible introduction, together with theoretical motivations to this Lévy market, can be found in, for example, Geman (2002). Some theoretical motivation for considering Lévy processes in finance can also be found in Leblanc and Yor (1998).

Note that since the law of X_t is infinitely divisible, it can be expressed for every n as the sum of n independent identically distributed random variables, with the law of $X_{t/n}$ as a common law. This is to be compared with the widely cited motivation for modelling stock returns by the Gaussian distribution, namely, that this distribution is a limiting distribution of sums of n independent random variables (up to a scaling factor), which may be viewed as representing the effects of various shocks in the economy.

Diffusion Component

Geman et al. (2001) have suggested that while price processes for financial assets must have a jump component they need not have a diffusion component. Jumps are necessary in order to capture the large moves that occasionally occur. The explanation usually given for the use of a diffusion component is that it captures the small moves which occur much more frequently. However, most of the above-mentioned pure jump models are infinite activity Lévy processes, i.e. with $\int_{-\infty}^{+\infty} \nu(dx) = \infty$, and they are able to capture both rare large moves and frequent small moves. High activity is accounted for by a large (in most cases infinite) number of small jumps. The empirical performance of these models is typically not improved by adding a diffusion component for returns. Thus, we mainly focus on the case where we take for X a pure jump process, i.e. with no Brownian component ($\sigma = 0$).

6.2.1 Market Incompleteness

From the PRP for Lévy processes (see Chapter 5), we see that, except when X is a Poisson process or a Brownian motion, our Lévy market model is an incomplete model. Note that the Poissonian case makes no sense economically, since a Poisson process has only up-jumps of size one.

The predictable integrands $a^{(i)}$, $i = 1, 2, \ldots$, appearing in the PRP can be obtained explicitly by solving PDEs (see Nualart and Schoutens 2001) or by using Mallivan calculus (see Benth *et al.* 2003; Di Nunno 2001; Di Nunno *et al.* 2002; Løkka 2001; Øksendal and Proske 2002). Applying the representation for a contingent claim, these $a^{(i)}$ can be interpreted in terms of (minimal-variance) hedging strategies. The processes $a^{(i)}$, $i = 2, 3, \ldots$, together correspond to the risk that cannot be hedged away. The term $a^{(1)}$ leads to the strategy that realizes the 'closest' hedge to the claim.

Leon *et al.* (2002) approximate the Lévy process by a sum of a Brownian motion and a countable number of compensated Poisson processes (see Section 8.2). They then introduce enough additional securities to complete the market. Via Malliavin calculus and more precisely by applying the Clark–Ocone–Haussman formula, they calculate the hedging portfolio in the approximated market.

In our Lévy market there are many different equivalent martingales measures to choose. In general this leads to many different possible prices for European options. Eberlein and Jacod (1997) prove that, for models based on an infinite-variation Lévy process, the range of call option prices that can be calculated in this way is the whole no-arbitrage interval. The boundaries of this interval are given by the condition that if the price lies beyond either of these boundaries, there is a simple buy/sell-and-hold strategy that allows a riskless arbitrage.

6.2.2 The Equivalent Martingale Measure

We focus on two ways to find an equivalent martingale measure.

The Esscher Transform

Following Gerber and Shiu (1994, 1996), we can by using the so-called Esscher transform find in some cases at least one equivalent martingale measure Q.

Let $f_t(x)$ be the density of our model's (real world, i.e. under P) distribution of X_t. For some real number $\theta \in \{\theta \in \mathbb{R} \mid \int_{-\infty}^{+\infty} \exp(\theta y) f_t(y) \, dy < \infty\}$ we can define a new density

$$f_t^{(\theta)}(x) = \frac{\exp(\theta x) f_t(x)}{\int_{-\infty}^{+\infty} \exp(\theta y) f_t(y) \, dy}. \tag{6.1}$$

Now we choose θ such that the discounted price process $\{\exp(-(r - q)t)S_t, t \geqslant 0\}$ is a martingale, i.e.

$$S_0 = \exp(-(r - q)t)E^{(\theta)}[S_t], \tag{6.2}$$

where expectation is taken with respect to the law with density $f_t^{(\theta)}(x)$. Let $\phi(u) = E[\exp(uiX_1)]$ denote the characteristic function of X_1. Then from (6.2) it can be shown that in order to let the discounted price process be a martingale, we need to have

$$\exp(r - q) = \frac{\phi(-i(\theta + 1))}{\phi(-i\theta)}. \tag{6.3}$$

The solution of this equation, θ^* say, gives us the Esscher transform martingale measure through the density function $f_t^{(\theta^*)}(x)$.

The choice of the Esscher measure among the set of possible other equivalent martingale measures may be justified by a utility-maximizing argument (see Gerber and Shiu 1996). For a discussion of the Esscher transform for specific classes of semi-martingales with applications in finance and insurance, see Bühlmann *et al.* (1996).

Finally, we note that if ϕ is the characteristic function and $[\gamma, \sigma^2, \nu(dx)]$ the Lévy triplet of X_1, then the characteristic function $\phi^{(\theta)}$ of the Esscher transformed measure is given by

$$\log \phi^{(\theta)}(u) = \log \phi(u - i\theta) - \log \phi(-i\theta).$$

Moreover, this law remains infinitely divisible and its Lévy triplet

$$[\gamma^{(\theta)}, (\sigma^{(\theta)})^2, \nu^{(\theta)}(dx)]$$

is given by

$$\gamma^{(\theta)} = \gamma + \sigma^2\theta + \int_{-1}^{1} (\exp(\theta x) - 1)\nu(dx),$$

$$\sigma^{(\theta)} = \sigma,$$

$$\nu^{(\theta)}(dx) = \exp(\theta x)\nu(dx).$$

Examples

The Normal distribution.   First we note that in the Black–Scholes world the historical measure of the log returns over a period of length 1 follows a Normal$(\mu - \frac{1}{2}\sigma^2, \sigma^2)$ law and thus in this case $\phi(u) = \exp(iu(\mu - \frac{1}{2}\sigma^2) - \sigma^2 u^2/2)$. So, (6.3) becomes

$$r - q = \mu - \frac{1}{2}\sigma^2 + \frac{1}{2}\sigma^2(2\theta + 1),$$

or

$$\theta^* = (r - q - \mu)/\sigma^2.$$

The density function in the risk-neutral world is then given by

$$f_1^{(\theta^*)}(x) = f^{(\theta^*)}(x) = \frac{\exp(\theta^* x - (x - \mu + \frac{1}{2}\sigma^2)^2/(2\sigma^2))}{\int_{-\infty}^{+\infty} \exp(\theta^* y - (y - \mu + \frac{1}{2}\sigma^2)^2/(2\sigma^2))\,dy}.$$

After some obvious simplifications, it can be seen that this is the density function of a Normal$(r - q - \frac{1}{2}\sigma^2, \sigma^2)$ distribution, as could be expected from the Black–Scholes theory.

The Meixner distribution. If log returns (under P) follow a Meixner$(\alpha, \beta, \delta, m)$ law, then with the Esscher transform our equivalent martingale measure Q follows a Meixner$(\alpha, \alpha\theta^* + \beta, \delta, m)$ distribution (see Grigelionis 1999; Schoutens 2001), where θ^* is given by

$$\theta^* = \frac{-1}{\alpha}\left(\beta + 2\arctan\left(\frac{-\cos(\alpha/2) + \exp((m - r + q)/(2\delta))}{\sin(\alpha/2)}\right)\right).$$

The NIG distribution. If log returns under our historical (real world) measure P follow an NIG$(\alpha, \beta, \delta, m)$ law, (6.3) reduces to

$$r - q = m + \delta(\sqrt{\alpha^2 - (\beta + \theta)^2} - \sqrt{\alpha^2 - (\beta + \theta + 1)^2})$$

and our equivalent martingale measure Q follows an NIG$(\alpha, \theta^* + \beta, \delta, m)$ distribution.

The GH distribution. This case is rather complicated. However, the Esscher transform can be obtained numerically by applying Fourier inversion techniques. We refer to Prause (1999).

The Mean-Correcting Martingale Measure

Although the Esscher transform is sometimes easy to obtain, it is not clear that in reality the market chooses this kind of (exponential) transform. Another way to obtain an equivalent martingale measure Q is by mean correcting the exponential of a Lévy process. This can be done by the special parameter m, to which we devoted Section 5.4. First we estimate in some way all the parameters involved in the process; then we change the m parameter in an appropriate way such that the discounted stock-price process becomes a martingale. Recall that in the Black–Scholes model the mean $\mu - \frac{1}{2}\sigma^2$ (i.e. the m_{old} parameter) of the Normal distribution was changed into $r - q - \frac{1}{2}\sigma^2$ (the m_{new} parameter). Here we do exactly the same; we take the m_{new} parameter to be

$$m_{\text{new}} = m_{\text{old}} + r - q - \log\phi(-i),$$

where $\phi(x)$ is the characteristic function of the log return involving the m_{old} parameter. Note that in the Black–Scholes model $\log\phi(-i) = \mu$. This choice of m_{new} will imply that our discounted stock price $\tilde{S} = \{\tilde{S}_t = \exp(-(r - q)t)S_t, t \geqslant 0\}$ is a martingale.

In Table 6.2, we list the m_{new} parameters (i.e. leading to a risk-neutral setting) for the different models.

Note also that we can proceed as follows. First estimate in some way the parameters of the process with the m parameter fixed at 0. Then introduce, as in Chapter 5, a parameter m as a function of the estimated parameters (see Table 6.2).

Table 6.2 The m parameter for the mean-correcting equivalent martingale measure.

Model	m_{new}
CGMY	$r - q - C\Gamma(-Y)((M-1)^Y - M^Y + (G+1)^Y - G^Y)$
VG	$r - q + C\log((M-1)(G+1)/(MG))$
NIG	$r - q + \delta(\sqrt{\alpha^2 - (\beta+1)^2} - \sqrt{\alpha^2 - \beta^2})$
Meixner	$r - q - 2\delta(\log(\cos(\beta/2)) - \log(\cos((\alpha+\beta)/2)))$
GH	$r - q - \log\left(\left(\dfrac{\alpha^2 - \beta^2}{\alpha^2 - (\beta+1)^2}\right)^{v/2} \dfrac{K_v(\delta\sqrt{\alpha^2 - (\beta+1)^2})}{K_v(\delta\sqrt{\alpha^2 - \beta^2})}\right)$

6.2.3 Pricing Formulas for European Options

Given our market model, we focus now on the pricing of European options for which the payoff function is only a function of the terminal stock price, i.e. the stock price S_T at maturity T: $G(S_T)$ denotes the payoff of the derivative at its time of expiry T. Write $F(X_T) = G(S_0 \exp(X_T))$. In the case of the European call with strike price K, we have $G(S_T) = (S_T - K)^+$ and $F(X_T) = (S_0 \exp(X_T) - K)^+$.

Pricing Through the Density Function

For a European call option with strike price K and time to expiration T, the value at time 0 is therefore given by the expectation of the payoff under the martingale measure Q:

$$E_Q[\exp(-rT)\max\{S_T - K, 0\}].$$

If we take for Q the Esscher transform equivalent martingale measure, this expectation can be written as

$$\exp(-qT)S_0 \int_c^\infty f_T^{(\theta^*+1)}(x)\,dx - \exp(-rT)K \int_c^\infty f_T^{(\theta^*)}(x)\,dx, \tag{6.4}$$

where $c = \ln(K/S_0)$. Similar formulas can be derived for other derivatives with a payoff function, $G(S_T) = G(S_0 \exp(X_T)) = F(X_T)$, which depends only on the terminal value at time $t = T$.

Pricing through the Lévy Characteristics

In all cases where the underlying process is a Lévy process (for simplicity without a Brownian component) in the risk-neutral world and the price $V_t = V(t, X_t)$ at time t of a given derivative satisfies some regularity conditions (i.e. $V(t, x) \in C^{(1,2)}$), the function $V(t, x)$ can also be obtained by solving a partial differential integral equation

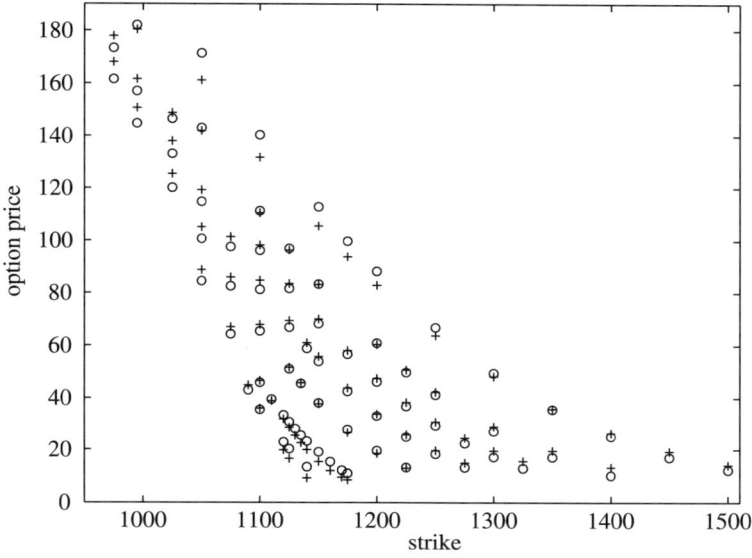

Figure 6.2   Meixner (mean-correcting) calibration of S&P 500 options
(circles are market prices, pluses are model prices).

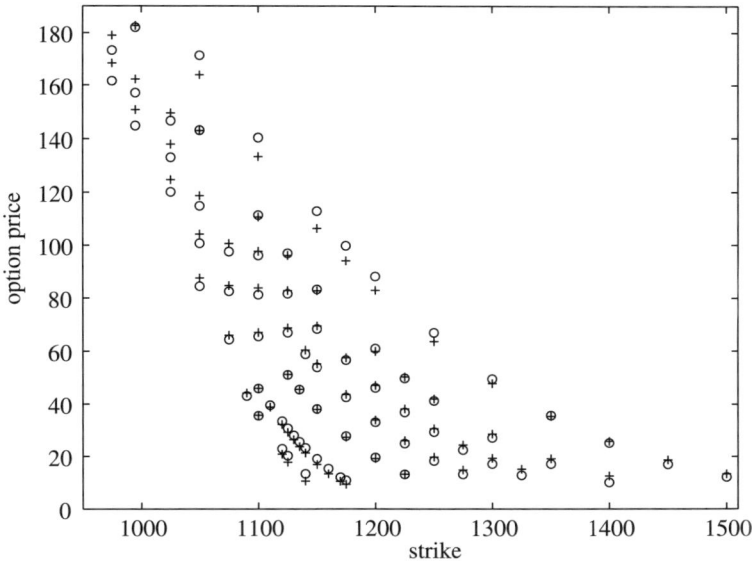

Figure 6.3   CGMY (mean-correcting) calibration of S&P 500 options
(circles are market prices, pluses are model prices).

Table 6.3 Lévy models (mean correcting): parameter estimation.

Model	Parameters			
	C	G	M	Y
CGMY	0.0244	0.0765	7.5515	1.2945
	α	β	δ	v
GH	3.8288	−3.8286	0.2375	−1.7555
	C	G	M	
VG	1.3574	5.8704	14.2699	
	α	β	δ	
NIG	6.1882	−3.8941	0.1622	
	α	β	δ	
Meixner	0.3977	−1.4940	0.3462	

(PDIE) with boundary condition, all in terms of the Lévy characteristics,

$$rV(t,x) = \gamma \frac{\partial}{\partial x} V(t,x) + \frac{\partial}{\partial t} V(t,x)$$
$$+ \int_{-\infty}^{+\infty} (V(t,x+y) - V(t,x) - y\frac{\partial}{\partial x} V(t,x))v^Q(dy),$$
$$V(T,x) = F(x),$$

where $[\gamma, 0, v^Q(dy)]$ is the triplet of Lévy characteristics of the Lévy process under the risk-neutral measure Q. This PDIE is the analogue of the famous Black–Scholes PDE (3.5) and follows from the Feynman–Kac formula for Lévy processes. It was derived in Nualart and Schoutens (2001) and Raible (2000).

Pricing through the Characteristic Function

In Chapter 2, the pricing method for European call options of Carr and Madan (1998) was given. It can be applied in general when the characteristic function of the risk-neutral log stock-price process is known.

Note that the expected value in $\varrho(v)$ of (2.4) is the characteristic function (in the value $(v - (\alpha + 1)i)$) of the logarithm of the stock price $\log(S_T)$ in the risk-neutral world at maturity. Since we are modelling our asset as an exponential of a Lévy process, we need only the characteristic function of our Lévy process at time T.

6.3 Calibration of Market Option Prices

The parameters which should resemble the market's view on the asset can be found through a calibration procedure on the market's option prices themselves. Here we do not explicitly take into account any historical data. All necessary information is contained in today's option prices, which we observe in the market. We estimate the

Table 6.4 Lévy models (mean-correcting): APE, AAE, RMSE, ARPE.

Model	APE (%)	AAE	RMSE	ARPE (%)
BS	8.87	5.4868	6.7335	16.92
CGMY	3.38	2.0880	2.7560	4.96
GH	3.60	2.2282	2.8808	5.46
VG	4.67	2.8862	3.5600	7.56
NIG	3.97	2.4568	3.1119	6.17
Meixner	4.19	2.5911	3.2451	6.71

model parameters by minimizing the root-mean-square error between the market's and the model's prices.

If we choose the mean-correcting equivalent martingale measure, we obtain a calibration for the Meixner and the CGMY models in Figures 6.2 and 6.3, respectively. In Table 6.3 we give the parameters coming from the calibration procedure. In Table 6.4 the relevant values of APE, AAE, RMSE and ARPE are given. Naturally, the four-parameter models perform better than the three-parameter models. Based on this calibration, however, it is not possible to say that one model is significantly better than the other. This calibration is only momentary and we typically see that calibrations to other datasets (of different underliers or on different times in history) can favour the model that performs worse here.

We see an improvement over the Black–Scholes prices. However, we still observe a significant difference from real market prices. It is typical that Lévy models incorporate by themselves a smile effect as in Figure 4.5 (see, for example, Eberlein *et al.* 1998; Schoutens 2001), although the effect does not completely correspond with the market.

7

Lévy Models with Stochastic Volatility

The main feature missing from the Lévy models described above is the fact that volatility (or more generally the environment) is changing stochastically over time. It has been observed that the estimated volatilities (or more generally the parameters of uncertainty) change stochastically over time and are clustered (see Section 4.2).

There are at least two ways of incorporating a volatility effect. The first method makes the volatility parameter of the Black–Scholes model stochastic in a suitable way. This technique has been implemented by Hull and White (1988) and Heston (1993). Their volatility process is driven by a Brownian motion. We will focus on other candidates for the stochastic description of this volatility parameter: OU processes for which the BDLP is a subordinator. This direction originates from a series of papers by Barndorff-Nielsen, Shephard and co-workers (Barndorff-Nielsen and Shephard 2001a,b, 2003b; Barndorff-Nielsen *et al.* 2002).

Another way of incorporating a similar effect was proposed by Carr, Madan, Geman and Yor. In Carr *et al.* (2003) they proposed the following. Increase or decrease the level of uncertainty by speeding up or slowing down the rate at which time passes. In order to build clustering and to keep time going forward, employ a mean-reverting positive process as a measure of the local rate of time change. The basic intuition underlying this approach arises from the Brownian motion scaling property (see Section 3.2.2), which relates changes in scale to changes in time. So, random changes in volatility can alternatively be captured by random changes in time. Candidates for the rate of time change are again the OU processes or the classical CIR process.

7.1 The BNS Model

In this section we investigate stochastic volatility extensions of the Black–Scholes model; the volatility follows an OU process driven by a subordinator. These models are called BNS models after Barndorff-Nielsen and Shephard. Sometimes, we simply refer to them as Black–Scholes-SV models.

Using Itô's lemma, we can transform the SDE for the stock-price process $S = \{S_t, t \geqslant 0\}$ under the Black–Scholes model,

$$dS_t = S_t(\mu\, dt + \sigma\, d\bar{W}_t), \quad S_0 > 0,$$

into an SDE equation for the log stock price $Z_t = \log(S_t)$:

$$dZ_t = d\log(S_t) = (\mu - \tfrac{1}{2}\sigma^2)\, dt + \sigma\, d\bar{W}_t, \quad \log S_0 = Z_0 = x_0.$$

If we want to take into account the fact that the volatility, i.e. σ, can change over time in an uncertain way, we can do this by making this parameter stochastic. Thus we will look for a stochastic process $\sigma^2 = \{\sigma_t^2, t \geqslant 0\}$ describing the nervousness of the market through time.

Moreover, it is often observed that a down-jump in the stock-price corresponds to an up-jump in volatility. In order to incorporate such a leverage effect, we impose a dependency structure.

Of particular interest is the model for σ^2 when σ^2 is an OU process (or a superposition of such processes (see Barndorff-Nielsen 2001)). Such models were introduced in this context by Barndorff-Nielsen and Shephard (2001a). In this case σ^2 satisfies an SDE of the form

$$d\sigma_t^2 = -\lambda\sigma_t^2\, dt + d\bar{z}_{\lambda t}, \tag{7.1}$$

where $z = \{\bar{z}_t, t \geqslant 0\}$ is a Lévy process with positive increments (a subordinator). We assume that $\bar{z}$ has no drift and its Lévy measure has a density. Letting

$$\hat{\theta} = \sup\{\theta \in \mathbb{R} : \log E[\exp(\theta\bar{z}_1)] < +\infty\},$$

we assume that

$$\hat{\theta} > 0 \quad \text{and} \quad \lim_{\theta \to \hat{\theta}} \log E[\exp(\theta\bar{z}_1)] = +\infty.$$

The log stock-price process now follows the dynamics,

$$dZ_t = d\log(S_t) = (\mu - \tfrac{1}{2}\sigma_t^2)\, dt + \sigma_t\, d\bar{W}_t + \rho\, d\bar{z}_{\lambda t}, \quad \log S_0 = Z_0 = x_0,$$

where ρ is a non-positive real parameter which accounts for the positive leverage effect. The Brownian motion and the BDLP are independent and we take as filtration, $\mathbb{F}$, the usual augmentation of the filtration generated by the pair $(\bar{W}, \bar{z})$.

As is typical for the more advanced market models, the model is arbitrage free but incomplete, which means that there exists more than one equivalent martingale measure. The structure of a general equivalent martingale measure and some relevant subsets are studied in Nicolato and Venardos (2003). Of special interest is the (structure-preserving) subset of martingale measures under which log returns are again described by a BNS model, albeit with different parameters and possibly different stationary laws, and as such a different law for the increments of the BDLP. Nicolato and Venardos (2003) argue that it is sufficient to consider only equivalent martingale measures of this subset. Barndorff-Nielsen et al. (2002) show that the dynamics of the log price under such an equivalent martingale measure Q are given by

$$dZ_t = (r - q - \lambda k(-\rho) - \tfrac{1}{2}\sigma_t^2)\, dt + \sigma_t\, dW_t + \rho\, dz_{\lambda t}, \quad Z_0 = x_0,$$
$$d\sigma_t^2 = -\lambda\sigma_t^2\, dt + dz_{\lambda t},$$

where $W = \{W_t, t \geq 0\}$ is a Brownian motion under Q independent of the BDLP $z = \{z_t, t \geq 0\}$ with the cumulant function of z_1 under Q given by $k(u) = k^Q(u) = \log E_Q[\exp(-uz_1)]$.

Under these dynamics, we can write the characteristic function of the log price in the form:

$$
\begin{aligned}
&\phi(u; t, \rho, \lambda, S_0, \sigma_0^2) \\
&= E_Q[\exp(iu \log S_t) \mid S_0, \sigma_0] \\
&= \exp(iu(\log(S_0) + (r - q - \lambda k^Q(-\rho))t) - \tfrac{1}{2}\lambda^{-1}(u^2 + iu)(1 - \exp(-\lambda t))\sigma_0^2)) \\
&\quad \times \exp\left(\lambda \int_0^t k(-\rho iu + \tfrac{1}{2}\lambda^{-1}(u^2 + iu)(1 - \exp(-\lambda(t - s))))\, ds\right).
\end{aligned}
$$

For the special choice of a Gamma–OU process or an IG–OU process, the expressions involved can be computed in terms of elementary functions. We set

$$f_1 = f_1(u) = iu\rho - \tfrac{1}{2}(u^2 + iu)(1 - \exp(-\lambda t)),$$
$$f_2 = f_2(u) = iu\rho - \tfrac{1}{2}(u^2 + iu).$$

7.1.1 The BNS Model with Gamma SV

The Gamma(a, b) distribution is self-decomposable (see Section 5.2.2). Thus there exists an OU process $\sigma^2 = \{\sigma_t^2, t \geq 0\}$ following the dynamics of the SDE (7.1) with a marginal Gamma(a, b) law. The corresponding BDLP z has Lévy density $w(x) = ab\exp(-bx)$; the associated cumulant function is given by $k(u) = -au(b + u)^{-1}$. Moreover, as stated in Barndorff-Nielsen *et al.* (2002), we can derive

$$
\begin{aligned}
\phi(u; t, \rho, \lambda, a, b, S_0, \sigma_0^2) &= E_Q[\exp(iu \log S_t) \mid S_0, \sigma_0] \\
&= \exp(iu(\log(S_0) + (r - q - a\lambda\rho(b - \rho)^{-1})t)) \\
&\quad \times \exp(-\tfrac{1}{2}\lambda^{-1}(u^2 + iu)(1 - \exp(-\lambda t))\sigma_0^2) \\
&\quad \times \exp\left(a(b - f_2)^{-1}\left(b\log\left(\frac{b - f_1}{b - iu\rho}\right) + f_2\lambda t\right)\right).
\end{aligned}
$$

The above characteristic function of the log stock price is exactly what is needed in the option-pricing formula (2.3) of Carr and Madan. We calibrate this model to our set of S&P 500 market option prices in Section 7.4.1.

7.1.2 The BNS Model with IG SV

The IG(a, b) distribution is also self-decomposable (see Section 5.2.2), so an OU process $\sigma^2 = \{\sigma_t^2, t \geqslant 0\}$ following the dynamics of the SDE (7.1) with a marginal IG(a, b) law exists. The corresponding BDLP z has a cumulant function given by $k(u) = -uab^{-1}(1 + 2ub^{-2})^{-1/2}$. Moreover, as stated in Barndorff-Nielsen *et al.* (2002), we can derive

$$\phi(u; t, \rho, \lambda, a, b, S_0, \sigma_0^2)$$

$$= E_Q[\exp(iu \log S_t) \mid S_0, \sigma_0]$$

$$= \exp(iu(\log(S_0) + (r - q - \rho\lambda ab^{-1}(1 - 2\rho b^{-2})^{-1/2})t))$$

$$\times \exp(\tfrac{1}{2}\lambda^{-1}(-u^2 - iu)(1 - \exp(-\lambda t))\sigma_0^2)$$

$$\times \exp(a(\sqrt{b^2 - 2f_1} - \sqrt{b^2 - 2iu\rho}))$$

$$\times \exp\left(\frac{2af_2}{\sqrt{2f_2 - b^2}}\left(\arctan\left(\sqrt{\frac{b^2 - 2iu\rho}{2f_2 - b^2}}\right) - \arctan\left(\sqrt{\frac{b^2 - 2f_1}{2f_2 - b^2}}\right)\right)\right).$$

We find the calibration of this model on our dataset of option prices on the S&P 500 Index in Section 7.4.1.

7.2 The Stochastic Time Change

The second way to build in stochastic volatility effects is to make time stochastic. In periods of high volatility, time will run faster than in periods of low volatility. Thus, it is possible that under a high volatility regime the return over one calendar day will equal the return over several days counted in the stochastic business time. In periods of lower volatility, one calendar day can correspond to only a part of a business day. Thus, cumulative effects can give rise to higher stock returns when volatility is high, and, similarly, smaller returns when volatility is low.

The application of stochastic time change to asset pricing goes back to Clark (1973), who modelled the asset price as a geometric Brownian motion subordinated by an independent Lévy subordinator.

In this section we give an overview of the variety of possible stochastic processes which can serve for the rate of time change. Since time needs to increase, all processes modelling the rate of time change need to be positive. The first candidate is the classical mean-reverting CIR process, which is based on Brownian motion. A second group of candidates is the OU processes driven by a subordinator. Examples of such processes are the Gamma–OU process and the IG–OU process. These processes will give rise to analytically tractable formulas in the context of option pricing.

7.2.1   The Integrated CIR Time Change

The CIR Process

Carr *et al.* (2003) use as the rate of time change the classical example of a mean-reverting positive stochastic process: the Cox–Ingersoll–Ross (CIR) process $y = \{y_t, t \geqslant 0\}$ (see Cox *et al.* 1985) that solves the SDE,

$$dy_t = \kappa(\eta - y_t)\,dt + \lambda y_t^{1/2}\,dW_t,$$

where $W = \{W_t, t \geqslant 0\}$ is a standard Brownian motion. The parameter η is interpreted as the long-run rate of time change, κ is the rate of mean reversion, and λ governs the volatility of the time change.

Note that, for $c > 0$, $\tilde{y} = cy = \{cy_t, t \geqslant 0\}$ satisfies the SDE

$$d\tilde{y}_t = \kappa(c\eta - \tilde{y}_t)\,dt + \sqrt{c}\lambda\tilde{y}_t^{1/2}\,dW_t, \tag{7.2}$$

and the initial condition is $\tilde{y}_0 = cy_0$.

The mean and variance of y_t given y_0 are given by

$$E[y_t \mid y_0] = y_0 \exp(-\kappa t) + \eta(1 - \exp(-\kappa t)),$$
$$\mathrm{var}[y_t \mid y_0] = y_0 \frac{\lambda^2}{\kappa}(\exp(-\kappa t) - \exp(-2\kappa t)) + \frac{\eta\lambda^2}{2\kappa}(1 - \exp(-\kappa t))^2.$$

The Integrated CIR Process

The economic time elapsed in t units of calendar time is then given by the integrated CIR process, $Y = \{Y_t, t \geqslant 0\}$, where

$$Y_t = \int_0^t y_s\,ds.$$

Since y is a positive process, Y is an increasing process.

The characteristic function of Y_t (given y_0) is explicitly known (see Cox *et al.* (1985) or Elliot and Kopp (1999, Theorem 9.6.3)),

$$E[\exp(iuY_t) \mid y_0] = \varphi(u, t; \kappa, \eta, \lambda, y_0)$$
$$= \frac{\exp(\kappa^2\eta t/\lambda^2)\exp(2y_0iu/(\kappa + \gamma\coth(\gamma t/2)))}{(\cosh(\gamma t/2) + \kappa\sinh(\gamma t/2)/\gamma)^{2\kappa\eta/\lambda^2}},$$

where

$$\gamma = \sqrt{\kappa^2 - 2\lambda^2 iu}.$$

From this we can derive

$$E[Y_t \mid y_0] = \eta t + \kappa^{-1}(y_0 - \eta)(1 - \exp(-\kappa t)).$$

7.2.2   The IntOU Time Change

Another possible choice for the rate of time change is an OU process driven by a subordinator (see Section 5.2.2). This kind of time change gives rise to jumps in the rate of time change process. Most of the time, volatility jumps up when new information is released; after the up-jump, it tends to gradually decrease. We consider the Gamma–OU case and the IG–OU case.

The rate of time change is now a solution of an SDE of the form

$$\mathrm{d}y_t = -\lambda y_t \, \mathrm{d}t + \mathrm{d}z_{\lambda t}, \tag{7.3}$$

where the process $z = \{z_t, t \geqslant 0\}$ is (since time has to increase) a subordinator, i.e. a nondecreasing Lévy process.

The economic time elapsed in t units of calendar time is then given by the corresponding intOU process, $Y = \{Y_t, t \geqslant 0\}$, where

$$Y_t = \int_0^t y_s \, \mathrm{d}s.$$

The Gamma–OU Time Change

The Gamma(a, b) distribution is self-decomposable (see Section 5.2.2). There thus exists an OU process $y = \{y_t, t \geqslant 0\}$ following the dynamics of the SDE (7.3) with a marginal Gamma(a, b) law.

Moreover, in the Gamma–OU case the characteristic function of Y_t (given y_0) can be given explicitly:

$$\begin{aligned}
\varphi_{\text{Gamma–OU}}&(u; t, \lambda, a, b, y_0) \\
&= E[\exp(\mathrm{i}u Y_t) \mid y_0] \\
&= \exp\left(\mathrm{i}u y_0 \lambda^{-1}(1 - \exp(-\lambda t)) \right. \\
&\qquad \left. + \frac{\lambda a}{\mathrm{i}u - \lambda b}\left(b\log\left(\frac{b}{b - \mathrm{i}u\lambda^{-1}(1 - \exp(-\lambda t))}\right) - \mathrm{i}ut \right) \right).
\end{aligned}$$

Note that we have

$$E[Y_t \mid y_0] = \lambda^{-1}(1 - \exp(-\lambda t))y_0 + \lambda^{-1}(a/b)(\lambda t - 1 - \exp(-\lambda t)).$$

Since $E[Y_t \mid y_0]/t$ converges to a/b as $t \to \infty$, the fraction a/b can be seen as the long-run average rate of time change.

Remark 7.1. Note that, for $c > 0$, $\tilde{Y} = cY = \{cY_t, t \geqslant 0\}$ is a time change based on a Gamma–OU process with parameters a and b/c and initial condition $\tilde{y}_0 = cy_0$. This can be seen from the scaling property of the Gamma distribution or from the above characteristic function.

IG–OU Time Change

As mentioned in Section 5.2.2, the IG law is also self-decomposable. There thus exists a stationary process $y = \{y_t, t \geq 0\}$ following the dynamics of the SDE (7.3) with a marginal $IG(a, b)$ law.

Recall that in this case the characteristic function of Y_t is explicitly known and given by

$$\varphi_{\text{IG–OU}}(u; t, \lambda, a, b, y_0) = E[\exp(iuY_t) \mid y_0]$$
$$= \exp\left(\frac{iuy_0}{\lambda}(1 - \exp(-\lambda t)) + \frac{2aiu}{b\lambda}A(u, t)\right),$$

where

$$A(u, t) = \frac{1 - \sqrt{1 + \kappa(1 - \exp(-\lambda t))}}{\kappa}$$
$$+ \frac{1}{\sqrt{1 + \kappa}}\left(\operatorname{artanh}\left(\frac{\sqrt{1 + \kappa(1 - \exp(-\lambda t))}}{\sqrt{1 + \kappa}}\right) - \operatorname{artanh}\left(\frac{1}{\sqrt{1 + \kappa}}\right)\right),$$
$$\kappa = -2b^{-2}iu/\lambda.$$

Again the fraction a/b can be seen as the long-run average rate of time change.

Remark 7.2. Note that, for $c > 0$, $\tilde{Y} = cY = \{cY_t, t \geq 0\}$ is a time change based on an IG–OU process with parameters $a\sqrt{c}$ and $b/\sqrt{c}$ and initial condition $\tilde{y}_0 = cy_0$. This can be seen from the scaling property of the IG distribution.

7.3 The Lévy SV Market Model

Let $Y = \{Y_t, t \geq 0\}$ be the process we choose to model our business time, i.e. our time change. Let us denote by $\varphi(u; t, y_0)$ the characteristic function of Y_t given y_0.

The (risk-neutral) price process $S = \{S_t, t \geq 0\}$ is now modelled as follows,

$$S_t = S_0 \frac{\exp((r - q)t)}{E[\exp(X_{Y_t}) \mid y_0]} \exp(X_{Y_t}),$$

where $X = \{X_t, t \geq 0\}$ is a Lévy process with

$$E[\exp(iuX_t)] = \exp(t\psi_X(u));$$

$\psi_X(u)$ is the characteristic exponent of the Lévy process. The factor

$$\frac{\exp((r - q)t)}{E[\exp(X_{Y_t}) \mid y_0]}$$

puts us immediately into the risk-neutral world by a mean-correcting argument.

Basically, we model the stock-price process as the ordinary exponential of a time-changed Lévy process. The process incorporates jumps (through the Lévy process)

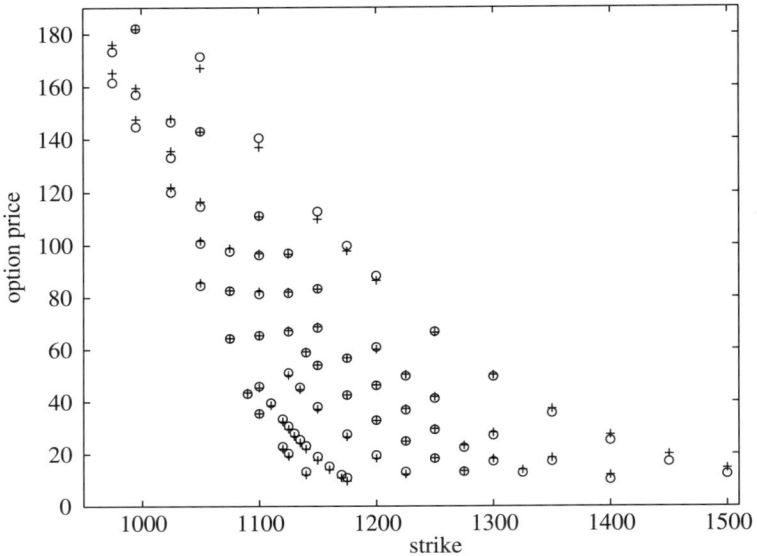

Figure 7.1   BNS OU–Gamma calibration of S&P 500 options
(circles are market prices, pluses are model prices).

and stochastic volatility (through the time change). Empirical work has generally
supported the need for both ingredients. Stochastic volatility appears to be needed to
explain the variation in strike of option prices at longer terms, while jumps are needed
to explain the variation in strike at shorter terms.

The characteristic function $\phi(u) = \phi(u; t, S_0, y_0)$ for the log of our stock price is
given by

$$\phi(u) = E[\exp(iu \log(S_t)) \mid S_0, y_0]$$
$$= \exp(iu((r - q)t + \log S_0)) \frac{\varphi(-i\psi_X(u); t, y_0)}{\varphi(-i\psi_X(-i); t, y_0)^{iu}}. \tag{7.4}$$

The characteristic function is important for the pricing of vanilla options (see formula
(2.3)). Recall that in these methods we only needed the characteristic function of
$\log(S_t)$. By the above formula, explicit formulas are thus at hand.

Note that if our Lévy process $X = \{X_t, t \geqslant 0\}$ is a VG or a CGMY process, for
$c > 0$, $\tilde{X} = \{X_{ct}, t \geqslant 0\}$ is again a Lévy process of the same class, with the same
parameters except the C parameter, which is multiplied now by the constant c. The
same can be said for the NIG and the Meixner processes. The parameter which takes
into account the same time-scaling property is now the δ parameter. In combination
with (7.2), Remarks 7.1 and 7.2, this means that in these cases there is one redundant
parameter. We therefore can set $y_0 = 1$ and scale the present rate of time change to 1.

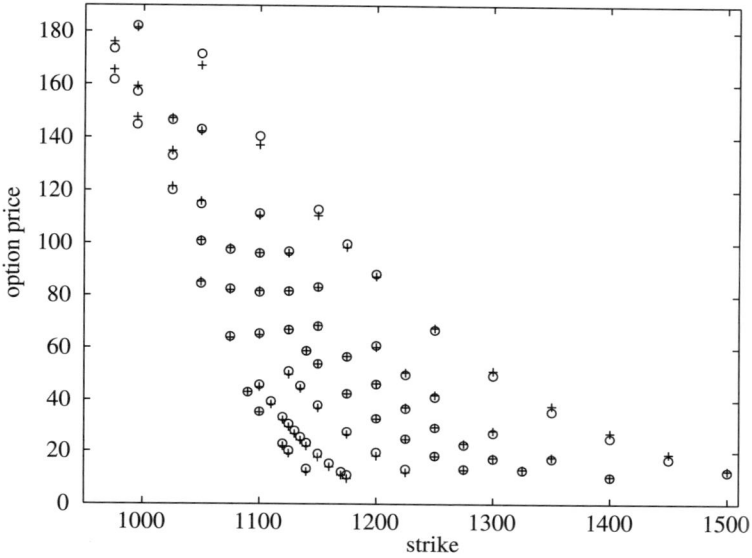

Figure 7.2   BNS OU–IG calibration of S&P 500 options
(circles are market prices, pluses are model prices).

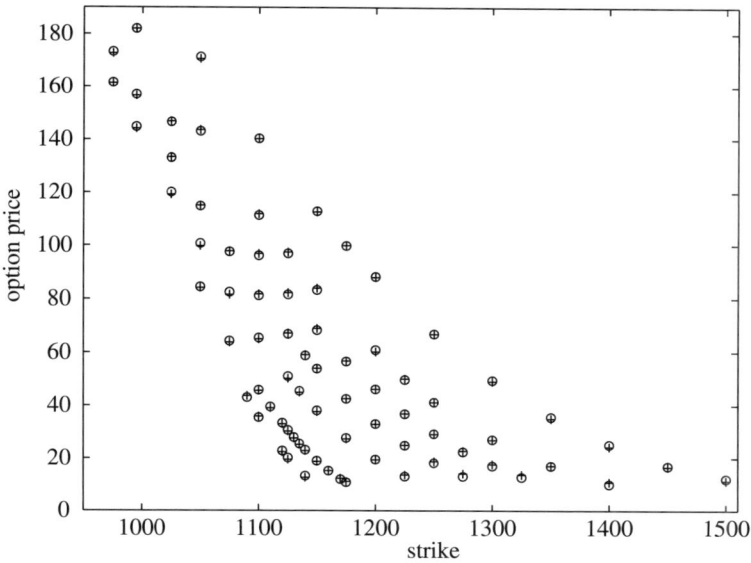

Figure 7.3   Meixner–CIR calibration of S&P 500 options
(circles are market prices, pluses are model prices).

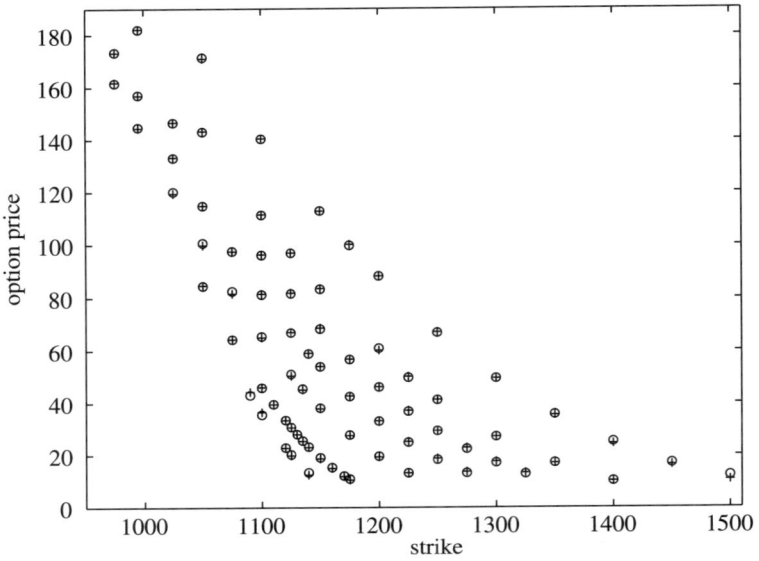

Figure 7.4   Meixner–OU–Gamma calibration of S&P 500 options
(circles are market prices, pluses are model prices).

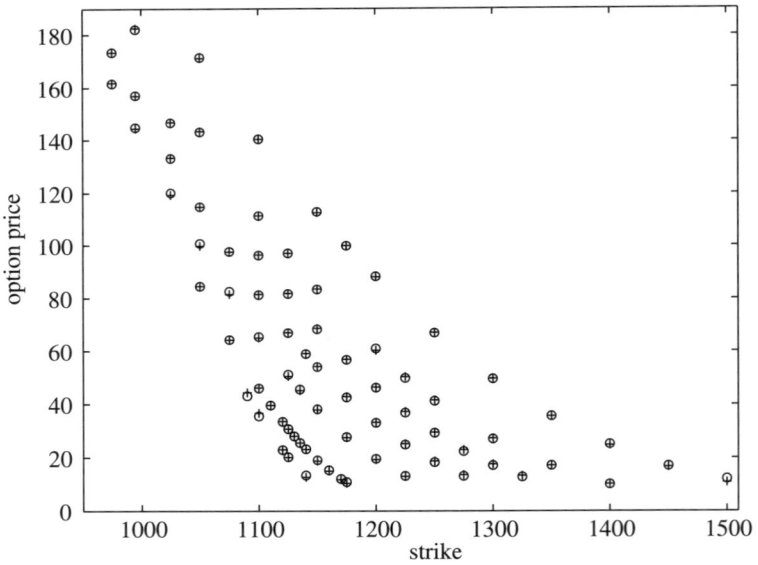

Figure 7.5   Meixner–OU–IG calibration of S&P 500 options
(circles are market prices, pluses are model prices).

Table 7.1 Parameter estimation.

BNS OU–Gamma				
ρ	λ	a	b	σ_0^2
-1.2606	0.5783	1.4338	11.6641	0.0145

BNS OU–IG				
ρ	λ	a	b	σ_0^2
-0.1926	0.0636	6.2410	0.7995	0.0156

Table 7.2 APE, AAE, RMSE and ARPE for BNS models.

Model	APE (%)	AAE	RMSE	ARPE (%)
BNS Gamma–OU	1.80	1.1111	1.4440	3.26
BNS IG–OU	1.62	1.0037	1.3225	2.71

More precisely, we have that the characteristic function $\phi(u)$ of (7.4) satisfies

$$\phi_{\text{CGMY--CIR}}(u; C, G, M, Y, \kappa, \eta, \lambda, y_0)$$
$$= \phi_{\text{CGMY--CIR}}(u; Cy_0, G, M, Y, \kappa, \eta/y_0, \lambda/\sqrt{y_0}, 1),$$
$$\phi_{\text{CGMY--Gamma--OU}}(u; C, G, M, Y, \lambda, a, b, y_0)$$
$$= \phi_{\text{CGMY--Gamma--OU}}(u; Cy_0, G, M, Y, \lambda, a, by_0, 1),$$
$$\phi_{\text{CGMY--IG--OU}}(u; C, G, M, Y, \lambda, a, b, y_0)$$
$$= \phi_{\text{CGMY--IG--OU}}(u; Cy_0, G, M, Y, \lambda, a/\sqrt{y_0}, b\sqrt{y_0}, 1),$$
$$\phi_{\text{VG--CIR}}(u; C, G, M, \kappa, \eta, \lambda, y_0)$$
$$= \phi_{\text{VG--CIR}}(u; Cy_0, G, M, \kappa, \eta/y_0, \lambda/\sqrt{y_0}, 1),$$
$$\phi_{\text{VG--Gamma--OU}}(u; C, G, M, \lambda, a, b, y_0)$$
$$= \phi_{\text{VG--Gamma--OU}}(u; Cy_0, G, M, \lambda, a, by_0, 1),$$
$$\phi_{\text{VG--IG--OU}}(u; C, G, M, \lambda, a, b, y_0)$$
$$= \phi_{\text{VG--IG--OU}}(u; Cy_0, G, M, \lambda, a/\sqrt{y_0}, b\sqrt{y_0}, 1),$$
$$\phi_{\text{Meixner--CIR}}(u; \alpha, \beta, \delta, \kappa, \eta, \lambda, y_0)$$
$$= \phi_{\text{Meixner--CIR}}(u; \alpha, \beta, \delta y_0, \kappa, \eta/y_0, \lambda/\sqrt{y_0}, 1),$$
$$\phi_{\text{Meixner--Gamma--OU}}(u; \alpha, \beta, \delta, \lambda, a, b, y_0)$$
$$= \phi_{\text{Meixner--Gamma--OU}}(u; \alpha, \beta, \delta y_0, \lambda, a, by_0, 1),$$
$$\phi_{\text{Meixner--IG--OU}}(u; \alpha, \beta, \delta, \lambda, a, b, y_0)$$
$$= \phi_{\text{Meixner--IG--OU}}(u; \alpha, \beta, \delta y_0, \lambda, a/\sqrt{y_0}, b\sqrt{y_0}, 1),$$

Table 7.3   Parameter estimation for Lévy SV models.

CGMY–CIR							
C	G	M	Y	κ	η	λ	y_0
0.0074	0.1025	11.3940	1.6765	0.3881	1.4012	1.3612	1
CGMY–Gamma–OU							
C	G	M	Y	λ	a	b	y_0
0.0415	3.9134	30.6322	1.3664	0.8826	0.5945	0.8524	1
CGMY–IG–OU							
C	G	M	Y	λ	a	b	y_0
0.0672	6.1316	44.7448	1.2911	1.0622	0.6092	0.9999	1
VG–CIR							
C	G	M		κ	η	λ	y_0
11.9896	25.8523	35.5344		0.6020	1.5560	1.9992	1
VG–Gamma–OU							
C	G	M		λ	a	b	y_0
11.4838	23.2880	40.1291		1.2517	0.5841	0.6282	1
VG–IG–OU							
C	G	M		λ	a	b	y_0
14.9248	26.1529	50.4425		1.2801	0.6615	0.8104	1
NIG–CIR							
α	β	δ		κ	η	λ	y_0
18.4815	−4.8412	0.4685		0.5391	1.5746	1.8772	1
NIG–Gamma–OU							
α	β	δ		λ	a	b	y_0
29.4722	−15.9048	0.5071		0.6252	0.4239	0.5962	1
NIG–IG–OU							
α	β	δ		λ	a	b	y_0
29.1553	−13.9331	0.5600		1.1559	0.6496	0.8572	1
Meixner–CIR							
α	β	δ		κ	η	λ	y_0
0.1231	−0.5875	3.3588		0.5705	1.5863	1.9592	1
Meixner–Gamma–OU							
α	β	δ		λ	a	b	y_0
0.1108	−0.9858	3.6288		1.1729	0.5914	0.6558	1

$\phi_{\text{NIG–CIR}}(u; \alpha, \beta, \delta, \kappa, \eta, \lambda, y_0)$
$$= \phi_{\text{NIG–CIR}}(u; \alpha, \beta, \delta y_0, \kappa, \eta/y_0, \lambda/\sqrt{y_0}, 1),$$
$\phi_{\text{NIG–Gamma–OU}}(u; \alpha, \beta, \delta, \lambda, a, b, y_0)$
$$= \phi_{\text{NIG–Gamma–OU}}(u; \alpha, \beta, \delta y_0, \lambda, a, b y_0, 1),$$
$\phi_{\text{NIG–IG–OU}}(u; \alpha, \beta, \delta, \lambda, a, b, y_0)$
$$= \phi_{\text{NIG–IG–OU}}(u; \alpha, \beta, \delta y_0, \lambda, a/\sqrt{y_0}, b\sqrt{y_0}, 1).$$

Table 7.3 *Cont.*

Meixner–IG–OU							
α	β	δ		λ	a	b	y_0
0.0890	−1.1323	5.0262		1.2190	0.6564	0.8266	1
GH–CIR							
α	β	δ	υ	κ	η	λ	y_0
8.3031	−4.8755	1.0297	−11.4534	0.5145	0.9029	1.3750	0.5748
GH–Gamma–OU							
α	β	δ	υ	λ	a	b	y_0
2.0267	−9.9780	0.3413	−5.6934	1.0740	0.3573	0.6143	2.0448
GH–IG–OU							
α	β	δ	υ	λ	a	b	y_0
23.2979	−12.9901	0.3463	−3.3960	1.1315	0.8993	0.6168	1.8345

Also, instead of setting the y_0 parameter equal to 1, the other parameters involved, e.g. δ or C, can be scaled to 1.

Actually, this time-scaling effect lies at the heart of the idea of incorporating stochastic volatility through making time stochastic. Here, it comes down to the fact that instead of making the volatility parameter (of the Black–Scholes model) stochastic, we are making the parameter C (in the VG and the CGMY cases), or the parameter δ (in the NIG and the Meixner cases), stochastic (via the time). Note that this effect not only influences the standard deviation (or volatility) of the processes; the skewness and the kurtosis are also now fluctuating stochastically.

7.4 Calibration of Market Option Prices

In this section we calibrate all the above models to our set of option prices on the S&P 500 Index. In order to calculate option prices we can make use of the pricing formula (2.3), which is based on the characteristic function of the log price process of the stock.

The model parameters can be estimated by minimizing the root-mean-square error between the market's prices on close and the model's option prices, and this over all strikes and maturities.

We will use the parameters coming out of this calibration procedure to price exotic options by Monte Carlo simulations in Section 9.3.

7.4.1 Calibration of the BNS Models

In Figures 7.1 and 7.2, we can see the result of the calibration procedure of the BNS model with a Gamma–OU process and an IG–OU process for the volatility behaviour.

Table 7.4   APE, AAE, RMSE and ARPE for Lévy SV models.

Model	APE (%)	AAE	RMSE	ARPE (%)
CGMY–CIR	0.56	0.3483	0.4367	1.15
CGMY–Gamma–OU	0.42	0.2576	0.3646	0.90
CGMY–IG–OU	0.44	0.2728	0.3736	0.90
VG–CIR	0.69	0.4269	0.5003	1.33
VG–Gamma–OU	0.51	0.3171	0.4393	1.10
VG–IG–OU	0.52	0.3188	0.4306	1.05
NIG–CIR	0.67	0.4123	0.4814	1.32
NIG–Gamma–OU	0.58	0.3559	0.4510	1.27
NIG–IG–OU	0.53	0.3277	0.4156	1.05
Meixner–CIR	0.68	0.4204	0.4896	1.34
Meixner–Gamma–OU	0.49	0.3033	0.4180	1.06
Meixner–IG–OU	0.50	0.3090	0.4140	1.03
GH–CIR	0.65	0.4032	0.4724	1.30
GH–Gamma–OU	0.45	0.2782	0.3837	0.95
GH–IG–OU	0.49	0.3041	0.3881	1.01

Table 7.1 gives an overview of the risk-neutral parameters coming out of the calibration procedure and Table 7.2 gives the corresponding APE, AAE, RMSE and ARPE.

7.4.2   Calibration of the Lévy SV Models

In Figures 7.3–7.5, we can see that the Meixner SV models give a very good fit to the empirical option prices of our S&P 500 dataset. Similar calibration results for the NIG, GH, VG and CGMY stochastic volatility models can be obtained.

Table 7.3 gives an overview of the risk-neutral parameters coming out of the calibration procedure and Table 7.4 gives the corresponding measures of fit, i.e. APE, AAE, RMSE and ARPE.

7.5   Conclusion

Different stochastic volatility models were formulated. Two methods of introducing stochastic volatility were discussed: we can make the volatility parameter of the Black–Scholes model stochastic (the direct approach), or we can incorporate a similar effect by making a stochastic time change. For all the models considered, closed-form expressions for the characteristic function of the log price process were given. All the models involved were calibrated to market option prices and were capable of adequately fitting option prices over a wide range of strikes and maturities. The models using a stochastic time change gave the best fits. However, these models depend on six to eight parameters. Given the fact that the models using a direct approach depend on only five parameters, these models also perform very well.

The calibrated process may be used for pricing standard options not included in the calibration or to detect serious mispricings. Moreover, the models proposed provide us with a relatively parsimonious representation of the surface of vanilla options. These almost perfect representations of the vanilla option surface lead also to interesting applications to the pricing of exotic options (see Chapter 9). In order to price these exotics, we first need to analyse ways to sample paths of all the ingredients of the stock-price process (see Chapter 8).

Finally, we note that the models considered, which can simultaneously explain both the statistical (historical) and the risk-neutral dynamics, are important for understanding the change of measure chosen by the market.

8

Simulation Techniques

In this chapter we look at possible simulation techniques for the processes encountered so far. We show how a Lévy process can be simulated based on a compound Poisson approximation. Also, we sample paths from the OU processes; the simulation in this case can be based on a series representation.

It is possible that for very specific processes, other techniques are available: this is, for example, the case if the Lévy process can be represented explicitly as a time change of a Brownian motion, and techniques are already available for simulating the 'simpler' subordinator (the time change). Sampling a path from the Lévy process can be done by sampling a path from the subordinator and a Brownian motion and then pursuing the time change.

Assume that we have random number generators at hand which can provide us with standard Normal (Normal$(0, 1)$) and Uniform$(0, 1)$ random numbers. We do not go into detail about the performance or the different types of such random number generators. Throughout this chapter we denote series of Normal$(0, 1)$ and Uniform$(0, 1)$ random numbers by $\{v_n, n = 1, 2, \dots\}$ and $\{u_n, n = 1, 2, \dots\}$, respectively.

8.1 Simulation of Basic Processes

In this section we give an overview of how to simulate some of the basic processes. Simulations of the more involved processes will rely on these basic processes. We thus look first at simulation of the standard Brownian motion and the Poisson process. A general reference for simulations of solutions of SDEs is Kloeden and Platen (1992). For Monte Carlo methods in finance, see Jäckel (2002).

8.1.1 Simulation of Standard Brownian Motion

Since the standard Brownian motion $W = \{W_t, t \geqslant 0\}$ has Normally distributed independent increments, simulation of it is easy. We discretize time by taking time steps of size Δt, which we assume to be very small. We will simulate the value of the

Lévy Processes in Finance   W. Schoutens
© 2003 John Wiley & Sons, Ltd   ISBN: 0-470-85156-2

Brownian motion at the time points $\{n\Delta t, n = 0, 1, \dots\}$. We have

$$W_0 = 0, \qquad W_{n\Delta t} = W_{(n-1)\Delta t} + \sqrt{\Delta t}\, v_n, \qquad n \geqslant 1,$$

where $\{v_n, n = 1, 2, \dots\}$ is a series of standard Normal random numbers. Figure 3.1 shows a path of a standard Brownian motion. We refer to Jäckel (2002) for the discussion of other simulation schemes and their use in the context of finance.

8.1.2   Simulation of a Poisson Process

The simulation of a Poisson process $N = \{N_t, t \geqslant 0\}$ with intensity parameter λ can be done in several different ways. We consider the method of exponential spacings and a classical method based on uniform random variates.

The Method of Exponential Spacings

The method of exponential spacings makes use of the fact that the inter-arrival times of the jumps of the Poisson process follow an Exponential $\text{Exp}(\lambda)$ distribution with mean λ^{-1}, i.e. a Gamma$(1, \lambda)$ distribution. An $\text{Exp}(\lambda)$ random number, e_n, can be obtained from a Uniform$(0, 1)$ random number, u_n, by

$$e_n = -\log(u_n)/\lambda.$$

Let

$$s_0 = 0, \qquad s_n = s_{n-1} + e_n, \qquad n = 1, 2, \dots,$$

then we can sample a path of the Poisson process N in the time points $\{n\Delta t, n = 0, 1, \dots\}$:

$$N_0 = 0, \qquad N_{n\Delta t} = \sup(k : s_k \leqslant n\Delta t), \qquad n \geqslant 1.$$

Uniform Method

If we need to simulate a Poisson process with intensity parameter $\lambda > 0$ up to a time point $T > 0$, we can also proceed as follows. First generate a random variate N which is Poisson(λT) distributed. Next, simulate N independent random uniform numbers $u_1, \dots, u_N$. Denote by $u_{(1)} < u_{(2)} < \cdots < u_{(N)}$ the order-statistics of this sequence. Then the jump points of the Poisson process are given by the points $Tu_{(1)}, \dots, Tu_{(N)}$, i.e. the Poisson process has a value 0 for time points $t < Tu_{(1)}$. At $t = Tu_{(1)}$ the process jumps to 1 and stays there until $t = Tu_{(2)}$, where it jumps to 2, etc.

A path of a Poisson process with parameter $\lambda = 25$ is shown in Figure 8.1.

8.2   Simulation of a Lévy Process

To simulate a Lévy process, we exploit the well-known compound Poisson approximation of this process. The name refers to the fact that we approximate the Lévy

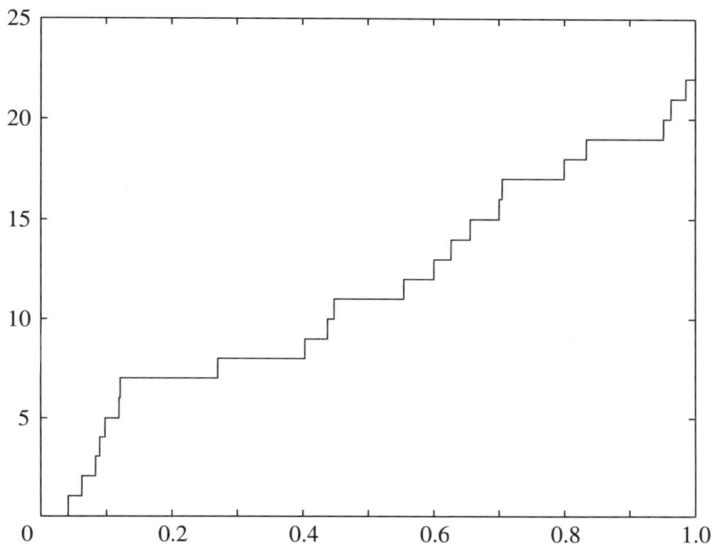

Figure 8.1   A sample path of a Poisson process.

process by a compound Poisson process. Special care has to be taken for the very small jumps. We can simply replace these very small jumps by their expected values. In some cases a further improvement can be made by replacing these small jumps by a Brownian motion. The procedure has been suggested on intuitive grounds and for some particular cases by, for example, Rydberg (1997b). Further support is given by Asmussen and Rosiński (2001) and Pollard (1984).

8.2.1   The Compound Poisson Approximation

General Procedure

The procedure is carried out as follows. Let X be a Lévy process with Lévy triplet $[\gamma, \sigma^2, \nu(dx)]$.

First, we discretize the Lévy measure $\nu(dx)$. We choose some small $0 < \epsilon < 1$. Then we make a partition of $\mathbb{R} \setminus [-\epsilon, \epsilon]$ of the following form. We choose real numbers

$$a_0 < a_1 < \cdots < a_k = -\epsilon, \quad \epsilon = a_{k+1} < a_{k+2} < \cdots < a_{d+1}.$$

Jumps larger than ϵ are approximated by a sum of independent Poisson processes in the following way. We take an independent Poisson process $N^{(i)} = \{N_t^{(i)}, t \geqslant 0\}$ for each interval, $[a_{i-1}, a_i), 1 \leqslant i \leqslant k$ and $[a_i, a_{i+1}), k + 1 \leqslant i \leqslant d$, with intensity λ_i given by the Lévy measure of the interval. Furthermore, we choose a point c_i (the jump size) in each interval such that the variance of the Poisson process matches the part of the variance of the Lévy process corresponding to this interval.

Approximation of the Small Jumps by Their Expected Value

Next, we look at the very small jumps. The first method is to simply replace them with their expected values.

This means that we approximate our Lévy process $X = \{X_t, t \geq 0\}$ by a process $X^{(d)} = \{X_t^{(d)}, t \geq 0\}$, which comprises a Brownian motion $W = \{W_t, t \geq 0\}$ and d independent Poisson processes $N^{(i)} = \{N_t^{(i)}, t \geq 0\}$, $i = 1, \ldots, d$, with intensity parameter λ_i:

$$X_t^{(d)} = \gamma t + \sigma W_t + \sum_{i=1}^{d} c_i (N_t^{(i)} - \lambda_i t 1_{|c_i| < 1}),$$

$$\lambda_i = \begin{cases} v([a_{i-1}, a_i)) & \text{for } 1 \leq i \leq k, \\ v([a_i, a_{i+1})) & \text{for } k+1 \leq i \leq d, \end{cases} \tag{8.1}$$

$$c_i^2 \lambda_i = \begin{cases} \displaystyle\int_{a_{i-1}}^{a_i -} x^2 v(\mathrm{d}x) & \text{for } 1 \leq i \leq k, \\ \displaystyle\int_{a_i}^{a_{i+1} -} x^2 v(\mathrm{d}x) & \text{for } k+1 \leq i \leq d. \end{cases} \tag{8.2}$$

If the original process has no Brownian component ($\sigma = 0$), then neither does the approximating process.

Approximation of the Small Jumps by a Brownian Motion

A further improvement is to also incorporate the contribution from the variation of small jumps. Write

$$\sigma^2(\epsilon) = \int_{|x| < \epsilon} x^2 v(\mathrm{d}x).$$

We let all (compensated) jumps smaller than ϵ contribute to the Brownian part of X. To be precise, we again approximate X by a process $X^{(d)}$, consisting of a Brownian motion $W = \{W_t, t \geq 0\}$ and d independent Poisson processes $N^{(i)} = \{N_t^{(i)}, t \geq 0\}$, $i = 1, \ldots, d$, with intensity parameter λ_i. Only the Brownian part is different from above. We now have

$$X_t^{(d)} = \gamma t + \tilde{\sigma} W_t + \sum_{i=1}^{d} c_i (N_t^{(i)} - \lambda_i 1_{|c_i| < 1} t),$$

where

$$\tilde{\sigma}^2 = \sigma^2 + \sigma^2(\epsilon),$$

and the λ_i and c_i, $i = 1, \ldots, d$, are as above in (8.1) and (8.2).

Note that a Brownian term appears even when the original process does not have one ($\sigma = 0$). In Asmussen and Rosiński (2001) a rigorous discussion is presented of when the latter approximation is valid. It turns out that this is the case if and only if for each $\kappa > 0$

$$\lim_{\epsilon \to 0} \frac{\sigma(\kappa\sigma(\epsilon) \wedge \epsilon)}{\sigma(\epsilon)} = 1. \tag{8.3}$$

This condition is implied by

$$\lim_{\epsilon \to 0} \frac{\sigma(\epsilon)}{\epsilon} = \infty. \tag{8.4}$$

Moreover, if the Lévy measure of the original Lévy process does not have atoms in some neighbourhood of the origin, then condition (8.4) and condition (8.3) are equivalent. Results on the speed of convergence of the above approximation can be found in Asmussen and Rosiński (2001).

Special Cases

The NIG Process. In Rydberg (1997b), the idea of replacing the small jumps by a Brownian motion was used for the NIG case. Computer simulations were provided to motivate the procedure. By condition (8.4), we can easily show that this is valid since in this case $\sigma(\epsilon) \sim \sqrt{2\alpha\delta/\pi}\epsilon^{1/2}$.

The Meixner Process. For the Meixner process we have also

$$\sigma(\epsilon) \sim \sqrt{2\alpha\delta/\pi}\epsilon^{1/2}.$$

Hence, $\sigma(\epsilon)/\epsilon \to \infty$ when $\epsilon \to 0$ and we can replace the small jumps by a Brownian component in the approximation.

The CGMY Process. Similarly, we can show for the CGMY process that $\sigma(\epsilon)/\epsilon \to \infty$ when $\epsilon \to 0$ only if $Y > 0$. Thus only for $Y > 0$ are we allowed to replace the small jumps by a Brownian component in the approximation.

The Gamma Process. Here we have that $\sigma(\epsilon)/\epsilon \to \sqrt{a/2}$ when $\epsilon \to 0$. Hence the approximation of small jumps by a Brownian motion fails.

The VG Process. Similarly, since a VG process is the difference of two Gamma processes, we cannot replace the small jumps by a Brownian component in the approximation. For an alternative approximation of the very small jumps, we refer to Emmer and Klüppelberg (2002).

8.2.2 On the Choice of the Poisson Processes

The choice of the intervals $[a_{i-1}, a_i), 1 \leqslant i \leqslant k$, and $[a_i, a_{i+1}), k+1 \leqslant i \leqslant d$, is crucial. For Lévy processes with a Lévy measure living on $\mathbb{R}$, we typically set $d = 2k$, so we have the same number of Poisson processes reflecting a positive as a negative jump. Next, we look at three different ways of choosing the intervals. First

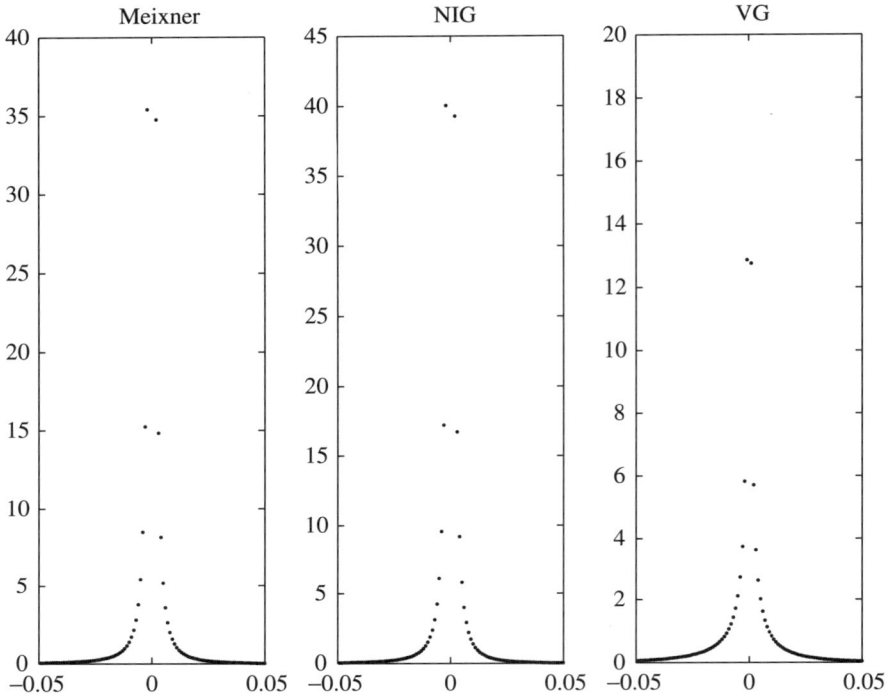

Figure 8.2 Equally spaced intervals.

we look at equally spaced intervals, then at equally weighted intervals, and finally at intervals with inverse linear boundaries. We illustrate this for the VG, NIG and Meixner processes, with $k = 100$ and parameters taken from the CIR combinations of Table 7.3.

Equally Spaced Intervals

We can choose the intervals to be equally spaced, i.e. $|a_{i-1} - a_i|$ is kept fixed for all $1 \leqslant i \leqslant d + 1, i \neq k + 1$. This choice is illustrated in Figure 8.2, where we plot λ_i versus c_i for all Lévy processes. A width equal to 0.001 was chosen and we zoomed in on the range $c_i \in [-0.05, 0.05]$; $k = 100$. Note the explosion near 0.

Equally Weighted Intervals

Here we opt to keep the intensities for the up-jumps and down-jumps corresponding to an interval constant. Thus, for equally weighted intervals, the Lévy measures of intervals on the negative part of the real line $v([a_{i-1} - a_i))$ are kept fixed for all $1 \leqslant i \leqslant k$. Similarly, the measure of intervals corresponding to up-jumps $v([a_i - a_{i+1}))$ is also kept fixed for all $k + 1 \leqslant i \leqslant d$. Note that for this choice the outer intervals can become quite large.

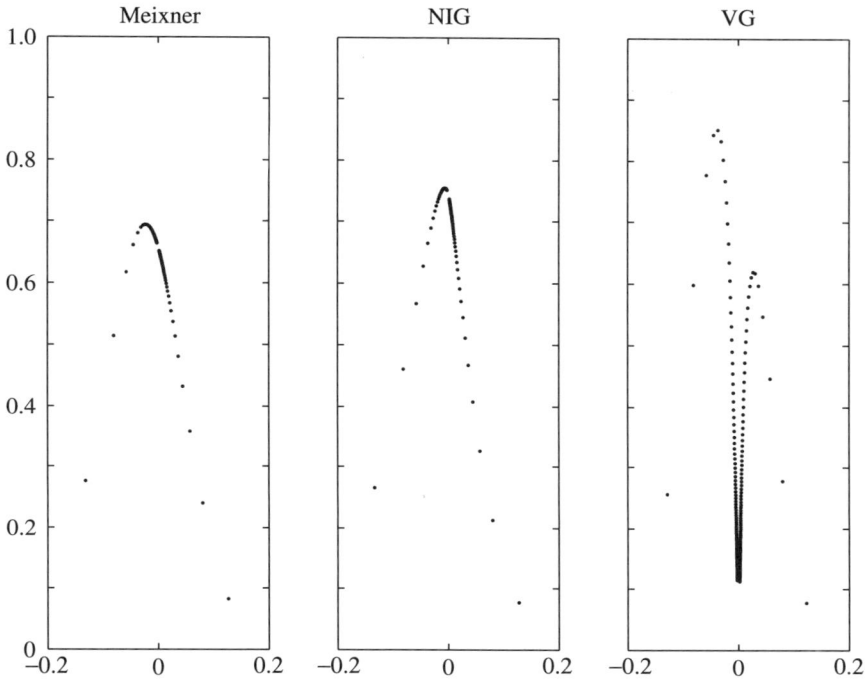

Figure 8.3 $a_{i-1} = -0.2/i$ and $a_{2k+2-i} = 0.2/i$, $1 \leqslant i \leqslant k+1$.

Interval with Inverse Linear Boundaries

Finally, we consider the case where the boundaries are given by $a_{i-1} = -\alpha i^{-1}$ and $a_{2k+2-i} = \alpha i^{-1}$, $1 \leqslant i \leqslant k+1$ and $\alpha > 0$. This leads to much more gradually decaying intensity parameters λ_i, as can be seen from Figure 8.3, where $\alpha = 0.2$ and $k = 100$. Moreover, there is no explosion to infinity near zero; the intensities even decrease again. Note that, in Figure 8.3, we now show the whole range with $c_i \in [-0.2, 0.2]$ for the same examples as above. Note also that in all cases the intensities of down-jumps are slightly higher than those of the corresponding up-jumps; this reflects the fact that log returns of stocks are negatively skewed.

8.3 Simulation of an OU Process

We will need to simulate from the process

$$y_t = \exp(-\lambda t)y_0 + \int_0^t \exp(-\lambda(t-s))\, dz_{\lambda s}$$

$$= \exp(-\lambda t)y_0 + \exp(-\lambda t)\int_0^{\lambda t} \exp(s)\, dz_s.$$

We will describe a method to do this by simulation directly from

$$\exp(-\lambda t) \int_0^{\lambda t} \exp(s)\, dz_s,$$

rather than (by the techniques of Section 8.2) from the BDLP $z = \{z_t, t \geqslant 0\}$. The idea is based on series representations. The required results can, in essence, be found in Marcus (1987) and Rosiński (1991). A self-contained overview is given in Barndorff-Nielsen and Shephard (2001b). Recent developments are surveyed in Rosiński (2001).

Let W be the Lévy measure of the BDLP z and let W^{-1} denote the inverse of the tail mass function W^+ as described in Section 5.2.

The crucial result is that in law

$$\int_0^t f(s)\, dz_s = \sum_{i=1}^{\infty} W^{-1}(a_i/t) f(t u_i), \tag{8.5}$$

where $\{a_i\}$ and $\{u_i\}$ are two independent sequences of random variables with u_i independent copies of a Uniform$(0, 1)$ random variable and $a_1 < \cdots < a_i < \cdots$ as the arrival times of a Poisson process with intensity 1.

It should be noted that the convergence of the series can be slow in some cases.

8.4   Simulation of Particular Processes

8.4.1   The Gamma Process

To simulate a Gamma process, we can use a Gamma random number generator.

Gamma Random Number Generators

First we note that, when X is Gamma(a, b), then, for $c > 0$, X/c is Gamma(a, bc). So we need only a good generator for Gamma$(a, 1)$ random numbers.

Next, we give two possible generators for Gamma$(a, 1)$ random numbers which should only be used for $a \leqslant 1$ (which is the case in all of our examples). The first one is called Johnk's Gamma generator (see Johnk 1964); the second one is Berman's Gamma generator (see Berman 1971).

Johnk's Gamma Generator

1. Generate two independent uniform random number u_1 and u_2.

2. Set $x = u_1^{1/a}$ and $y = u_2^{1/(1-a)}$.

3. If $x + y \leqslant 1$ goto step 4, else goto step 1.

4. Generate an Exp(1) random variable , i.e. $z = -\log(u)$, where u is a uniform random number.

5. Return the number $zx/(x + y)$ as the Gamma$(a, 1)$ random number.

Berman's Gamma Generator

1. Generate two independent uniform random number u_1 and u_2.

2. Set $x = u_1^{1/a}$ and $y = u_2^{1/(1-a)}$.

3. If $x + y \leqslant 1$ goto step 4, else goto step 1.

4. Generate two independent uniform random number u_1 and u_2.

5. Return the number $-x \log(u_1 u_2)$ as the Gamma$(a, 1)$ random number.

Several other generators are described in the literature. We refer to Devroye (1986) for a detailed survey.

Simulation of a Gamma Process

Next, it is easy to simulate a sample path of a Gamma process $G = \{G_t, t \geqslant 0\}$, where G_t follows a Gamma(at, b) law. We simulate the value of this process at time points $\{n\Delta t, n = 0, 1, \ldots\}$ as follows. First generate independent Gamma$(a\Delta t, b)$ random numbers $\{g_n, n \geqslant 1\}$ by, for example, the techniques described above. Note that since we assume Δt to be very small, $a\Delta t$ is in most cases smaller than 1 and we can use the Berman or Johnk generators. Then

$$G_0 = 0, \quad G_{n\Delta t} = G_{(n-1)\Delta t} + g_n, \quad n \geqslant 1.$$

Figure 8.4 shows a path of a Gamma process with parameters $a = 10$ and $b = 20$.

8.4.2   The VG Process

Simulation of a VG Process as the Difference of Two Gamma Processes

Since a VG process can be seen as the difference of two independent Gamma processes, simulation of a VG process is easy. More precisely, a VG process

$$X^{(VG)} = \{X_t^{(VG)}, t \geqslant 0\}$$

with parameters $C, G, M > 0$ can be decomposed as $X_t^{(VG)} = G_t^{(1)} - G_t^{(2)}$, where $G^{(1)} = \{G_t^{(1)}, t \geqslant 0\}$ is a Gamma process with parameters $a = C$ and $b = M$ and $G^{(2)} = \{G_t^{(2)}, t \geqslant 0\}$ is a Gamma process with parameters $a = C$ and $b = G$.

Figure 8.5 shows a path of a VG process with parameters $C = 20$, $G = 40$ and $M = 50$.

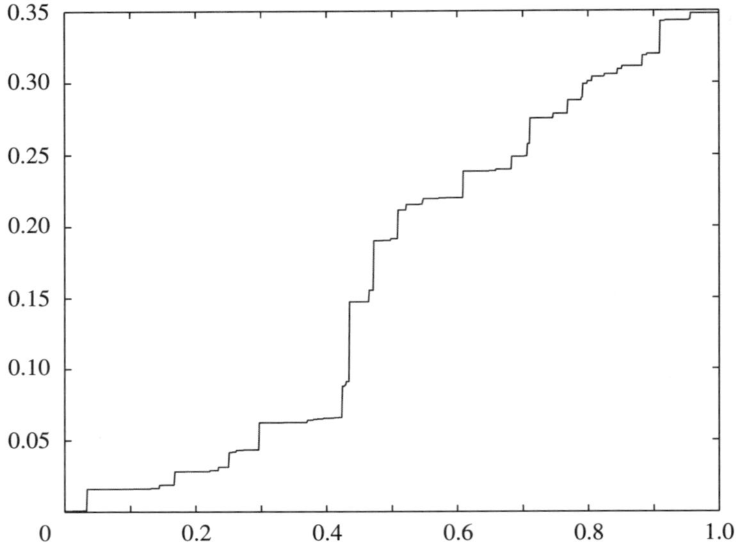

Figure 8.4   A sample path of a Gamma process.

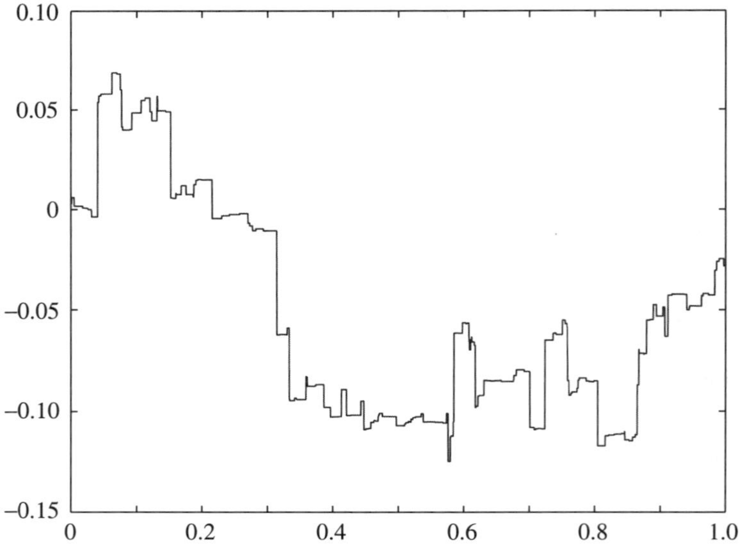

Figure 8.5   A sample path of a VG process.

Simulation of a VG Process as a Time-Changed Brownian Motion

We can also simulate a VG process as a time-changed Brownian motion. This proce-
dure can be best explained in the (σ, ν, θ) parametrization, rather than in the (C, G, M)

parametrization. Recall that a VG process $X^{(\mathrm{VG})} = \{X_t^{(\mathrm{VG})}, t \geqslant 0\}$ with parameters $\sigma > 0$, $\nu > 0$ and θ can be obtained by time-changing a standard Brownian motion $W = \{W_t, t \geqslant 0\}$ with drift by a Gamma process $G = \{G_t, t \geqslant 0\}$ with parameters $a = 1/\nu$ and $b = 1/\nu$. We have

$$X_t^{(\mathrm{VG})} = \theta G_t + \sigma W_{G_t}.$$

A sample path of the VG process can thus be obtained by sampling a standard Brownian motion and a Gamma process.

8.4.3 The TS Process

For the TS distribution, neither the density function nor specific random number generators are available. In order to simulate, we have to rely on other techniques. Rosiński (2001) (see also Rosiński 2002) describes a method based on the so-called rejection method. We approximate the path of a TS process $X = \{X_t, 0 \leqslant t \leqslant T\}$ with parameters $a > 0$, $b \geqslant 0$ and $0 < \kappa < 1$ by

$$X_t^{(K)} = \sum_{k=1}^{K} \min\left(2\left(\frac{aT}{b_i \Gamma(1-\kappa)}\right)^{1/\kappa}, \frac{2e_i \tilde{u}_i^{1/\kappa}}{b^{1/\kappa}}\right) 1_{(Tu_i < t)}, \quad 0 \leqslant t \leqslant T,$$

where $\{e_n, n = 1, 2, \dots\}$ is a sequence of independent $\mathrm{Exp}(1)$ random numbers, $\{u_n, n = 1, 2, \dots\}$, $\{\tilde{u}_n, n = 1, 2, \dots\}$ are sequences of independent Uniform$(0, 1)$ random numbers and $b_1 < b_2 < \cdots < b_i < \cdots$ are the arrival times of a Poisson process (independent of the other series) with intensity parameter 1. All series are assumed to be independent of each other.

Then, as $K \to \infty$, $X^{(K)} \to X$ uniformly (from below). We thus simulate the whole path directly. Note that the random numbers $\{e_n\}$, $\{u_n\}$, $\{\tilde{u}_n\}$ and $\{b_n\}$ are parameter free. An important aspect is the choice of K. As K increases we converge to the true path, from below. Typically, values of K around $10\,000$ give very reasonable approximations.

8.4.4 The IG Process

An IG Random Number Generator

To simulate an IG process, we can use the IG random number generator proposed by Michael *et al.* (1976) (see also Devroye 1986). In order to sample from an IG(a, b) distribution, we follow the following algorithm.

IG generator of Michael, Schucany and Haas.

1. Generate a standard Normal random number v.

2. Set $y = v^2$.

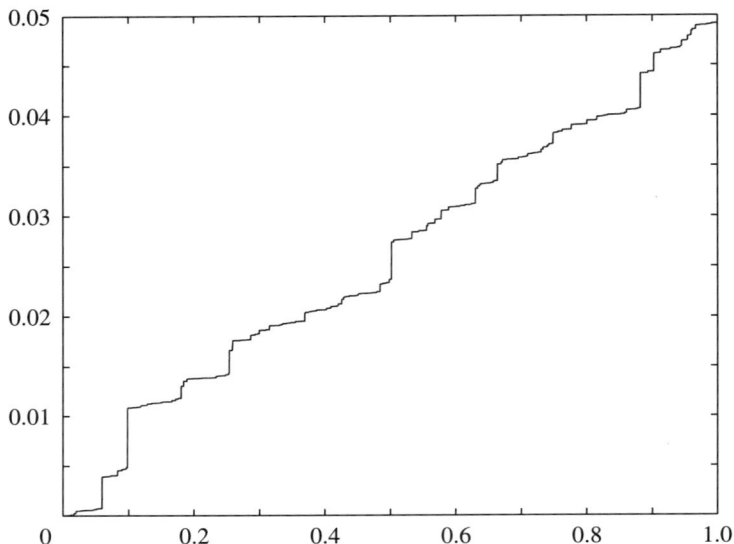

Figure 8.6 A sample path of an IG process.

3. Set $x = (a/b) + y/(2b^2) - \sqrt{4aby + y^2}/(2b^2)$.

4. Generate a uniform random number u.

5. If $u \leqslant a/(a + xb)$, then return the number x as the IG(a, b) random number, else return $a^2/(b^2 x)$ as the IG(a, b) random number.

Simulation of an IG Process using IG Random Numbers

Next, it is easy to simulate a sample path of an IG process $I = \{I_t, t \geqslant 0\}$, where I_t follows an IG(at, b) law. We simulate the value of this process at time points $\{n\Delta t, n = 0, 1, \ldots\}$ as follows. First generate independent IG$(a\Delta t, b)$ random numbers $\{i_n, n \geqslant 1\}$, then

$$I_0 = 0, \qquad I_{n\Delta t} = I_{(n-1)\Delta t} + i_n, \quad n \geqslant 1.$$

Figure 8.6 shows a path of an IG process with parameters $a = 1$ and $b = 20$.

Simulation of an IG Process by the Path Rejection Method

To simulate an IG process we can use the special technique of Section 8.4.3, which is valid for the more general TS processes. We approximate the path of an IG process $X = \{X_t, 0 \leqslant t \leqslant T\}$ with parameters $a > 0$ and $b > 0$ by

$$X_t^{(K)} = \sum_{k=1}^K \min\left(\frac{2}{\pi}\left(\frac{aT}{b_i}\right)^2, \frac{2e_i \tilde{u}_i^2}{b^2}\right) 1_{(Tu_i < t)}, \quad 0 \leqslant t \leqslant T,$$

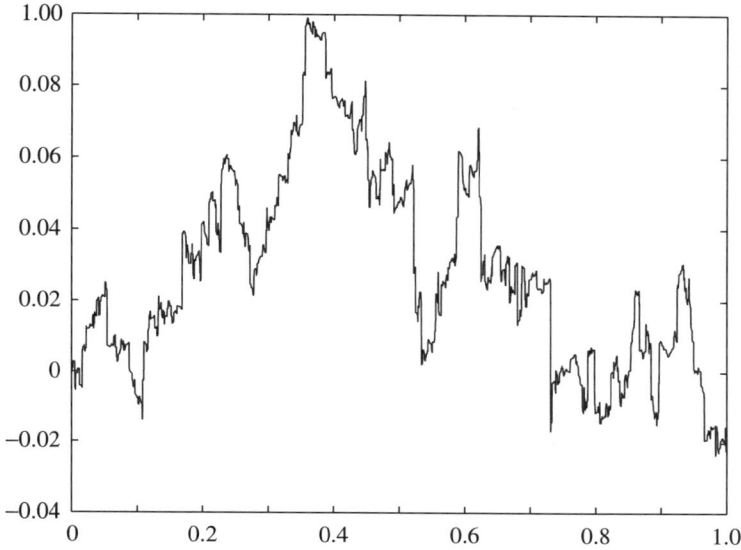

Figure 8.7 A sample path of an NIG process.

where $\{e_n, n = 1, 2, \ldots\}$ is a sequence of independent Exp(1) random numbers, $\{u_n, n = 1, 2, \ldots\}$, $\{\tilde{u}_n, n = 1, 2, \ldots\}$ are sequences of independent Uniform(0, 1) random numbers and $b_1 < b_2 < \cdots < b_i < \cdots$ are the arrival times of a Poisson process (independent of the other series) with intensity parameter 1. All series are assumed to be independent of each other.

Then, as $K \to \infty$, $X^{(K)} \to X$ uniformly (from below).

8.4.5 The NIG Process

Simulation of an NIG Process as a Time-Changed Brownian Motion

As in the VG case, we can also simulate an NIG process as a time-changed Brownian motion. Recall that an NIG process $X^{(\text{NIG})} = \{X_t^{(\text{NIG})}, t \geq 0\}$ with parameters $\alpha > 0$, $-\alpha < \beta < \alpha$ and $\delta > 0$ can be obtained by time-changing a standard Brownian motion $W = \{W_t, t \geq 0\}$ with drift by an IG process $I = \{I_t, t \geq 0\}$ with parameters $a = 1$ and $b = \delta\sqrt{\alpha^2 - \beta^2}$. We have

$$X_t^{(\text{NIG})} = \beta\delta^2 I_t + \delta W_{I_t}.$$

A sample path of the NIG process can thus be obtained by sampling a standard Brownian motion and an IG process.

Figure 8.7 shows a path of an NIG process with parameters $\alpha = 50$, $\beta = -10$ and $\delta = 1$.

8.4.6   The Gamma–OU Process

The Gamma–OU process can be simulated by the series representation described in the previous paragraph. The series in this case simplifies a lot. Another way to simulate a Gamma–OU process is via its BDLP.

By the Series Representation via the Inverse Tail Mass Function

The case of the Gamma–OU process with Gamma(a, b) marginals can benefit from the explicit expression of W^{-1} we have at hand:

$$W^{-1} = \max\{0, -b^{-1}\log(x/a)\}.$$

It can be shown that in law we can then rewrite (8.5) as

$$\exp(-\lambda t)\int_0^{\lambda t} \exp(s)\,\mathrm{d}z_s = b^{-1}\exp(-\lambda t)\sum_{i=1}^{N_1}\log(c_i^{-1})\exp(\lambda t u_i),$$

where $c_1 < c_2 < \cdots$ are the arrival times of a Poisson process $N = \{N_s, s \geqslant 0\}$ with intensity parameter $a\lambda t$ $(E[N_s] = a\lambda t s)$, N_1 as the corresponding number of events up until time 1 and as before the u_i independent copies of a Uniform$(0, 1)$ random variable.

Through the BDLP

The BDLP for the Gamma(a, b)–OU process $y = \{y_t, t \geqslant 0\}$ is a compound Poisson process, i.e. $z_t = \sum_{n=1}^{N_t} x_n$, where $N = \{N_t, t \geqslant 0\}$ is a Poisson process with intensity parameter a, i.e. $E[N_t] = at$ and $\{x_n, n = 1, 2, \ldots\}$ is an independent and identically distributed sequence; each x_n follows a Gamma$(1, b) = \mathrm{Exp}(b)$ law. The Poisson process can be simulated as described above. Recall that the exponential random numbers can be obtained from uniform random numbers: $x_n = -\log(u_n)/b$.

We base our simulation on the SDE,

$$\mathrm{d}y_t = -\lambda y_t\,\mathrm{d}t + \mathrm{d}z_{\lambda t}, \quad y_0 \geqslant 0.$$

To simulate a Gamma(a, b)–OU process $y = \{y_t, t \geqslant 0\}$ in the time points $t = n\Delta t, n = 0, 1, 2, \ldots$, first simulate in the same time points a Poisson process $N = \{N_t, t \geqslant 0\}$ with intensity parameter $a\lambda$, then (with the convention that an empty sum equals zero)

$$y_{n\Delta t} = (1 - \lambda\Delta t)y_{(n-1)\Delta t} + \sum_{n=N_{(n-1)\Delta t}+1}^{N_{n\Delta t}} x_n;$$

here the factor $(1 - \lambda\Delta t)$ can also be replaced by $\mathrm{e}^{-\lambda\Delta t}$.

Figure 8.8 shows a path of a Gamma–OU process with parameters $\lambda = 10, a = 10$, $b = 100$ and $y_0 = 0.08$. Note that the mean of the marginal Gamma(a, b) law is at $a/b = 0.1$.

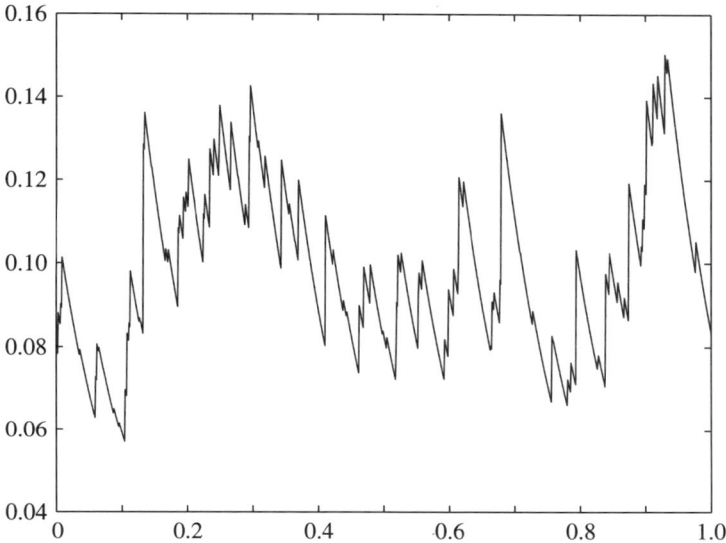

Figure 8.8   A sample path of a Gamma–OU process.

8.4.7   The IG–OU Process

The IG–OU process can be simulated by the series representation described in the previous paragraph. An approximation is also possible through special simulation techniques for the TS process; recall that the IG process is a special case of this TS process.

By the Series Representation via the Inverse Tail Mass Function

In the case of an IG–OU process with IG(a, b) marginals, we do not have an explicit expression of W^{-1} at hand. However, we can use an approximation:

$$W^{-1}(x) \sim \frac{a^2}{2\pi x^2}.$$

Plugging this approximation of the inverse of the tail integral of the BDLP into formula (8.5) gives us a way to simulate. However, the convergence in the series is rather slow and care needs to be taken with the truncation.

By the Series Representation for the TS Process

A special (rejection) method was developed by Rosiński (2001) (see also Rosiński 2002) to simulate paths in the tempered stable case. We make use of the fact that the BDLP can be decomposed into an IG process and a compound Poisson process as noted in Section 5.5.2. Recall that the (OU)–IG process is a special case of the (OU)–TS process ($\kappa = 1/2$).

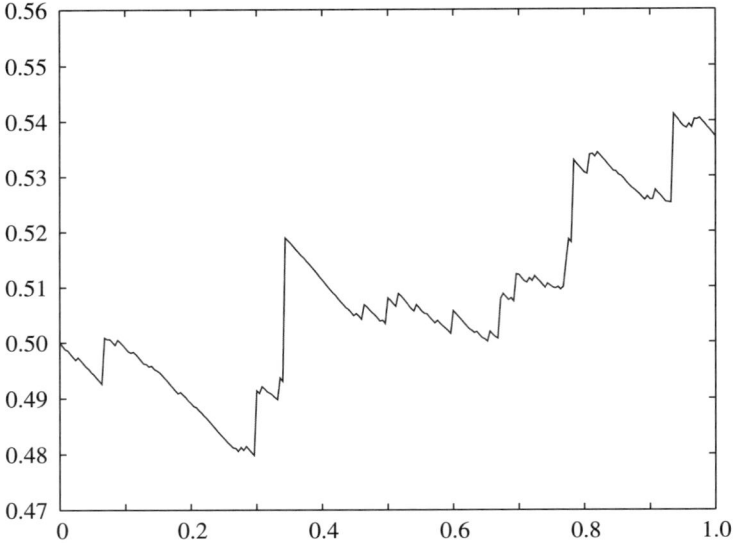

Figure 8.9 A sample path of an IG–OU process.

Let u_i, $\bar{u}_i$ and $\tilde{u}_i$ be sequences of independent uniform random numbers, let e_i be independent exponential random numbers with mean 1 and let $b_1 < b_2 < \cdots$ be the arrival times of a Poisson process with intensity 1; moreover, let $N = \{N_t, t \geqslant 0\}$ be a Poisson process with intensity parameter $ab/2$ and interarrival times $d_1 < d_2 < \cdots$, and v_i independent standard Normal random numbers. All sequences are independent of each other. The main result in the special situation of the IG–OU process is that

$$y_t = \exp(-\lambda t)y_0 + \exp(-\lambda t) \sum_{i=1}^{N_{\lambda t}} \frac{v_i^2}{b^2} \exp(d_i) + q_t^{(K)},$$

where $q_t^{(K)}$ is approximated in law by, for $0 \leqslant t \leqslant T$,

$$q_t^{(K)} = \exp(-\lambda t) \sum_{i=1}^{K} \min\left(\frac{1}{2\pi} \left(\frac{a\lambda T}{b_i} \right)^2, \frac{2e_i \tilde{u}_i^2}{b^2} \right) \exp(\lambda t \bar{u}_i) 1_{(T u_i \leqslant t)},$$

where the approximation error goes uniformly to zero as $K \to \infty$. $q_t^{(K)}$ converges to q_t from below.

The resulting picture typically becomes quite stable by the time K reaches 25 000. The advantage of this method is that it potentially represents the whole process as a function of t $(0 \leqslant t \leqslant T)$, and not just at a particular value of t. Figure 8.9 shows a path of an IG process with $K = 25\,000$, with parameters $\lambda = 0.3$, $a = 10$, $b = 20$ and $y_0 = 0.5$.

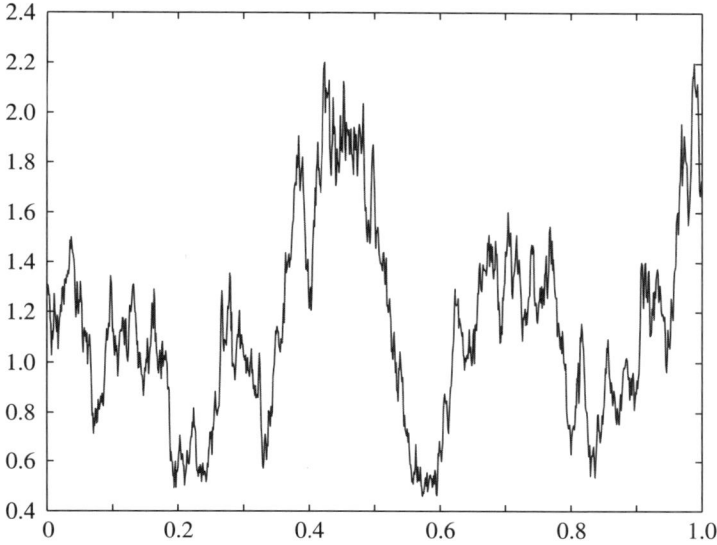

Figure 8.10 A sample path of a CIR process.

8.4.8 The CIR Process

The simulation of a CIR process $y = \{y_t, t \geqslant 0\}$ is quite easy and classical. Basically, we discretize the SDE,

$$dy_t = \kappa(\eta - y_t)\,dt + \lambda y_t^{1/2}\,dW_t, \quad y_0 \geqslant 0.$$

The sample path of the CIR process $y = \{y_t, t \geqslant 0\}$ in the time points $t = n\Delta t$, $n = 0, 1, 2, \ldots$, is then given by

$$y_{n\Delta t} = y_{(n-1)\Delta t} + \kappa(\eta - y_{(n-1)\Delta t})\Delta t + \lambda y_{(n-1)\Delta t}^{1/2}\sqrt{\Delta t}\,v_n,$$

where $\{v_n, n = 1, 2, \ldots\}$ is a series of independent standard normal random numbers.

Figure 8.10 shows a path of a CIR process with parameters $\kappa = 5$, $\eta = 1$, $\lambda = 2$ and $y_0 = 1.25$.

8.4.9 BNS Model

The log price process $Z = \{Z_t, t \geqslant 0\}$ of a stock under the BNS model follows the dynamics,

$$dZ_t = (\mu - \tfrac{1}{2}\sigma_t^2)\,dt + \sigma_t\,dW_t + \rho\,dz_{\lambda t}, \quad Z_0 = \log(S_0),$$
$$d\sigma_t^2 = -\lambda\sigma_t^2\,dt + dz_{\lambda t}, \quad \sigma_0^2 \geqslant 0,$$

where $\sigma^2 = \{\sigma_t^2, t \geqslant 0\}$ is an OU process driven by its BDLP $z = \{z_t, t \geqslant 0\}$, a subordinator. In order to sample the log price process, we need first to simulate the

OU process σ^2 in the points $t = n\Delta t$, $n = 0, 1, 2, \ldots$, together with the BDLP z in the time points $t = \lambda n\Delta t$, $n = 0, 1, 2, \ldots$. Then, sample a path of the log price process in the time points $t = n\Delta t$, $n = 0, 1, 2, \ldots$: $Z_0 = \log(S_0)$ and

$$Z_{n\Delta t} = Z_{(n-1)\Delta t} + (\mu - \sigma^2_{n\Delta t}/2)\Delta t + \sigma_{n\Delta t}\sqrt{\Delta t}\,v_n + \rho(z_{\lambda n\Delta t} - z_{\lambda(n-1)\Delta t}).$$

The simulation of the OU process and its BDLP can be done simultaneously with the simulation of the log price process. For each time point we first sample the values of the OU process and its BDLP and then we use these to find a sample value of the log price. Then we move on to the next time point. Note that by using the $\sigma^2_{n\Delta t}$ and $z_{\lambda n\Delta t}$ values to obtain the $Z_{n\Delta t}$ values, an up-jump in the OU process and its BDLP will give rise at the same moment to a jump in the stock price.

9

Exotic Option Pricing

The payoff of the option has in most cases until now depended only on the value of the underlying at expiry (see, for example, Section 2.5.2). However, path-dependent options have become popular in the OTC market in the last two decades. Examples of these exotic path-dependent options are lookback options and barrier options. Other exotics have no expiry date and the holder can exercise whenever he or she wants. These are called perpetual options.

In this chapter we first look at barrier and lookback options. We give their Black–Scholes prices, which are in closed form. Next, we indicate how one could in principle obtain prices in a Lévy market. However, the dimension of the problem is high and we need to make use of numerical inversion techniques and integrals with dimension three or four have to be calculated numerically.

We also give a survey of other exotics studied under a Lévy market in the literature. Most of them lead to prices only if some model restrictions are imposed.

Finally, we show how Monte Carlo techniques can give us an estimate of the prices of exotic options of European type. The Monte Carlo method is based on the simulation of the stock-price process as described in the previous chapter. In order to speed up the procedure we can make use of a variance-reduction technique based on control variates.

9.1 Barrier and Lookback Options

9.1.1 Introduction

The lookback call (put) option with floating strike has the particular feature of allowing its holder to buy (sell) the stock at the minimum (maximum) it has achieved over the life of the option. The payoff of a barrier option depends on whether the price of the underlying asset crosses a given threshold (the barrier) before maturity. The simplest barrier options are 'knock-in' options, which are activated when the price of the underlying asset touches the barrier, and 'knock-out' options, which are deactivated in that case. For example, an up-and-out call has the same payoff as a regular plain

Lévy Processes in Finance   W. Schoutens
© 2003 John Wiley & Sons, Ltd   ISBN: 0-470-85156-2

vanilla call if the price of the underlying asset remains below the barrier over the life of the option, but becomes worthless as soon as the price of the underlying asset crosses the barrier.

Let us consider contracts of duration T, and denote the maximum and minimum process, respectively, of a process $X = \{X_t, 0 \leqslant t \leqslant T\}$ by

$$M_t^X = \sup\{X_u; 0 \leqslant u \leqslant t\} \quad \text{and} \quad m_t^X = \inf\{X_u; 0 \leqslant u \leqslant t\}, \quad 0 \leqslant t \leqslant T.$$

Using risk-neutral valuation and choosing an equivalent martingale measure Q, we have that the initial, i.e. $t = 0$, price of a lookback call option is given by

$$LC = \exp(-rT)E_Q[S_T - m_T^S];$$

the initial price of a lookback put is given by

$$LP = \exp(-rT)E_Q[M_T^S - S_T].$$

Recall that the indicator function is denoted by $1(A)$, which has a value 1 if A is true and 0 otherwise.

For single-barrier options, we will focus on the following types of call options.

- The down-and-out barrier call is worthless unless its minimum remains above some low barrier H, in which case it retains the structure of a European call with strike K. Its initial price is given by

$$DOBC = \exp(-rT)E_Q[(S_T - K)^+ 1(m_T^S > H)].$$

- The down-and-in barrier call is a standard European call with strike K if its minimum goes below some low barrier H. If this barrier is never reached during the lifetime of the option, the option is worthless. Its initial price is given by

$$DIBC = \exp(-rT)E_Q[(S_T - K)^+ 1(m_T^S \leqslant H)].$$

- The up-and-in barrier call is worthless unless its maximum crosses some high barrier H, in which case it retains the structure of a European call with strike K. Its price is given by

$$UIBC = \exp(-rT)E_Q[(S_T - K)^+ 1(M_T^S \geqslant H)].$$

- The up-and-out barrier call is worthless unless its maximum remains below some high barrier H, in which case it retains the structure of a European call with strike K. Its price is given by

$$UOBC = \exp(-rT)E_Q[(S_T - K)^+ 1(M_T^S < H)].$$

The put counterparts, replacing $(S_T - K)^+$ with $(K - S_T)^+$, can be defined along the same lines.

We note that the value, DIBC, of the down-and-in barrier call option with barrier H and strike K plus the value, DOBC, of the down-and-out barrier option with the same barrier H and the same strike K is equal to the value C of the vanilla call with strike K. The same is true for the up-and-out together with the up-and-in:

$$
\left.
\begin{aligned}
\text{DIBC} + \text{DOBC} &= \exp(-rT) E_Q[(S_T - K)^+ (1(m_T^S \geqslant H) + 1(m_T^S < H))] \\
&= \exp(-rT) E_Q[(S_T - K)^+] \\
&= C, \\
\text{UIBC} + \text{UOBC} &= \exp(-rT) E_Q[(S_T - K)^+ (1(M_T^S \geqslant H) + 1(M_T^S < H))] \\
&= \exp(-rT) E_Q[(S_T - K)^+] \\
&= C.
\end{aligned}
\right\}
$$

(9.1)

9.1.2 Black–Scholes Barrier and Lookback Option Prices

Under the Black–Scholes framework, it is possible to obtain explicit expressions for the prices of the above-mentioned exotic options. For other exotics, see, for example, Haug (1998).

Barrier Options

We have that, if $H \leqslant K$,

$$
\begin{aligned}
\text{DIBC} &= S_0 \exp(-qT)(H/S_0)^{2\lambda} \text{N}(y) \\
&\quad - K \exp(-rT)(H/S_0)^{2\lambda-2} \text{N}(y - \sigma\sqrt{T}), \\
\text{DOBC} &= C - \text{DIBC}, \\
\text{UOBC} &= 0, \\
\text{UIBC} &= C,
\end{aligned}
$$

and, if $H > K$,

$$
\begin{aligned}
\text{DOBC} &= S_0 \text{N}(x_1) \exp(-qT) - K \exp(-rT) \text{N}(x_1 - \sigma\sqrt{T}) \\
&\quad - S_0 \exp(-qT)(H/S_0)^{2\lambda} \text{N}(y_1) \\
&\quad + K \exp(-rT)(H/S_0)^{2\lambda-2} \text{N}(y_1 - \sigma\sqrt{T}), \\
\text{DIBC} &= C - \text{DOBC}, \\
\text{UIBC} &= S_0 \text{N}(x_1) \exp(-qT) - K \exp(-rT) \text{N}(x_1 - \sigma\sqrt{T}) \\
&\quad - S_0 \exp(-qT)(H/S_0)^{2\lambda} (\text{N}(-y) - \text{N}(-y_1)) \\
&\quad + K \exp(-rT)(H/S_0)^{2\lambda-2} (\text{N}(-y + \sigma\sqrt{T}) - \text{N}(-y_1 + \sigma\sqrt{T})), \\
\text{UOBC} &= C - \text{UIBC},
\end{aligned}
$$

where

$$\lambda = \sigma^{-2}(r - q + \tfrac{1}{2}\sigma^2),$$
$$y = (\sigma\sqrt{T})^{-1}\log(H^2/(S_0 K)) + \lambda\sigma\sqrt{T},$$
$$x_1 = (\sigma\sqrt{T})^{-1}\log(S_0/H) + \lambda\sigma\sqrt{T},$$
$$y_1 = (\sigma\sqrt{T})^{-1}\log(H/S_0) + \lambda\sigma\sqrt{T}.$$

An important issue for barrier options is the frequency with which the stock price is observed for the purpose of determining whether the barrier has been reached. The above formulas assume a continuous observation. Often, the terms of the barrier contract are modified and there are only a discrete number of observations, for example, at the close of each trading day. Broadie $et\ al.$ (1997) provide a way of adjusting the above formulas for periodic observations. The barrier H is replaced by

$$H \exp(0.582\sigma\sqrt{T/m})$$

for an up-and-in or up-and-out option and by

$$H \exp(-0.582\sigma\sqrt{T/m})$$

for a down-and-in or down-and-out barrier, where m is the number of times the stock prices is observed; T/m is the time interval between observations.

Lookback Options

For the lookback options the following closed formulas are available for the initial price:

$$LC = S_0 e^{-qT}(N(a_1) - \sigma^2(2(r - q))^{-1}N(-a_1))$$
$$\qquad - S_0 e^{-rT}(N(a_2) - \sigma^2(2(r - q))^{-1}N(-a_2)),$$
$$LP = S_0 e^{-qT}(\sigma^2(2(r - q))^{-1}N(a_1) - N(-a_1))$$
$$\qquad + S_0 e^{-rT}(N(-a_2) - \sigma^2(2(r - q))^{-1}N(-a_2)),$$

where

$$a_1 = \sigma^{-1}(r - q + \tfrac{1}{2}\sigma^2)\sqrt{T},$$
$$a_2 = \sigma^{-1}(r - q - \tfrac{1}{2}\sigma^2)\sqrt{T}.$$

As with barrier options, the value of the lookback option is also liable to be sensitive to the frequency with which the asset price is observed for the computation of the maximum or minimum. The above formulas assume continuous observation. Broadie $et\ al.$ (1999) provide a way of adjusting the formulas when the observations are discrete.

9.1.3 Lookback and Barrier Options in a Lévy Market

Assume that we work in the Lévy market of Section 6.2 and that in the risk-neutral setting, i.e. under the equivalent martingale measure Q, we have that the stock price is the exponential of a Lévy process: $S_t = S_0 \exp(X_t)$. Assume for simplicity that X_t has for all $0 \leqslant t \leqslant T$ a density function (under Q) denoted by $f_t(x)$.

Finding explicit formulas for exotic options in the more sophisticated Lévy market is very hard.

Barrier options under a Lévy market were considered by Boyarchenko and Levendorskiǐ (2002c). The results rely on the Wiener–Hopf decomposition and analytic techniques are used. Similar and totally general results by probabilistic methods for barrier and lookback options are described by Yor and Nguyen (2001). The numerical calculations needed are highly complex: numerical integrals with dimension 3 or 4 are needed, together with numerical inversion methods.

We follow Yor and Nguyen (2001) and sketch how prices can in principle be calculated. We focus on the up-and-in barrier call option with payoff function:

$$(S_T - K)^+ 1(M_T^S \geqslant H). \tag{9.2}$$

Its initial price is denoted by UIBC = UIBC(S_0, T, K, H, r). The first step to note (by differentiating (9.2) with respect to K) that we can write (independent of the model used), for all K, T, r, S_0 and H,

$$\text{UIBC}(S_0, T, K, H, r) = \int_K^\infty \text{BUIC}(S_0, T, k, H, r) \, dk, \tag{9.3}$$

where BUIC(S_0, T, K, H, r) is the price of a binary up-and-in call with the same maturity, i.e. with payoff function at maturity

$$1(S_T \geqslant K) 1(M_T^S \geqslant H).$$

This option pays out one currency unit if the price of the stock at maturity is above the strike K only if during the lifetime of the option the stock price has risen above some barrier H. In all other cases the option expires worthless.

This option is a kind of barrier-version of the vanilla binary call option. A vanilla binary call option with maturity T and strike K has a payoff function given by

$$1(S_T \geqslant K)$$

and pays out one currency unit if the stock at expiry is above the strike K, and expires worthless otherwise. We will denote its price by BC(S_0, T, K, r). We have

$$\text{BC}(S_0, T, K, r) = \exp(-rT) E_Q[1_{S_T \geqslant K}]$$
$$= \exp(-rT) \int_{\log(K/S_0)}^\infty f_T(x) \, dx.$$

In order to avoid the integration (9.3) required to obtain the price of an up-and-in barrier call option from the price of a binary up-and-in call option, we can also use the Esscher transform, already encountered in Section 6.2.2. To do this we have to impose the condition that $E[\exp(X_1)] < \infty$. This hypothesis allows us to consider the Esscher transform of the distribution of X_T with density function $f_T(x)$. We define a new distribution with density function:

$$f_T^{(1)}(x) = \frac{\exp(x)f_T(x)}{\int_{-\infty}^{+\infty}\exp(y)f_T(y)\,dy}.$$

This is the Esscher transform of (6.1) for $\theta = 1$. Then, as noted in Section 6.2.2, X remains a Lévy process under this new measure. The characteristic function, $\phi^{(1)}$ of this new measure is now given in terms of the characteristic function, ϕ of the original measure by the relation:

$$\log\phi^{(1)}(u) = \log\phi(u - i) - \log\phi(-i).$$

We can rewrite the payoff function (9.2) of the up-and-in barrier as

$$(S_0\exp(X_T) - K)1(M_T^S \geqslant H)1(S_T \geqslant K).$$

Hence the price is given by

$$\text{UIBC} = (S_0\,\text{BUIC}^{(1)} - K\,\text{BUIC}).$$

Here BUIC denotes the price of the binary up-and-in call under the original setting, i.e. using ϕ as the characteristic function of the underlying Lévy process, and $\text{BUIC}^{(1)}$ denotes the price of the binary up-and-in call using $\phi^{(1)}$ as the characteristic function of the underlying Lévy process.

An important role will be played by the function

$$\kappa(\alpha, \beta) = \exp\left(\int_0^\infty\int_0^\infty \frac{\exp(-t) - \exp(-\alpha t - \beta x)}{t}f_t(x)\,dx\,dt\right),$$

where κ is (up to a constant) the Laplace exponent of the ladder process (see Bertoin 1996). Knowledge of this function is needed to apply the Pecherskii–Rogozin identity, which expresses the double Laplace transform of the joint distribution of hitting times and the value of the process at such times in terms of the function κ. This identity was first proved by Pecherskii and Rogozin (1969) using Wiener–Hopf analysis techniques.

Let $x > 0$ and define the first passage time of X above x to be

$$T(x) = T_X(x) = \inf\{t > 0 : X_t > x\}$$

and the so-called overshoot to be

$$K(x) = K_X(x) = X_{T_X(x)} - x.$$

Then, for every $\alpha, \beta, q > 0$, the following Pecherskii–Rogozin identity holds:

$$\int_0^\infty \exp(-ux)E[\exp(-\alpha T(x) - \beta K(x))]\,dx = \frac{\kappa(\alpha, u) - \kappa(\alpha, \beta)}{(u - \beta)\kappa(\alpha, u)}. \qquad (9.4)$$

Inversion of this (triple) Laplace transform gives rise to the joint distribution of the first passage time and the process its value at that time. This information is needed for the calculation of the binary up-and-in call price, BUIC. Let $\tau = T_X(\log(H/S_0))$. Then

$$\begin{aligned}
\text{BUIC}(S_0, T, K, H, r) &= \exp(-rT)E_Q[1_{S_T \geqslant K} 1_{M_T^S \geqslant H}] \\
&= \exp(-rT)E_Q[1_{X_T \geqslant \log(K/S_0)} 1_{T_X(\log(H/S_0)) \leqslant T}] \\
&= \exp(-rT)E_Q[E_Q[1_{X_T \geqslant \log(K/S_0)} 1_{\tau \leqslant T} \mid \mathcal{F}_\tau]] \\
&= E_Q[\exp(-r\tau)1_{\tau \leqslant T} \text{ BC}(S_0 \exp(X_\tau), T - \tau, K, r)].
\end{aligned}$$

In order to calculate the final expectation, the joint law of τ and X_τ is needed. This can be obtained by inversion of the Pecherskii–Rogozin identity.

Lookback options and the other types of barrier options can be treated along the same lines.

It is clear that the calculation under this market model is quite involved, and it is not clear whether the numerical calculation is in general better than the Monte Carlo techniques we will discuss in Section 9.3. Further simplification of the formulas (in special cases) is the subject of ongoing research and could possibly lead to a reduction of the computational drawbacks of the method. Some simplifications are available in Yor and Nguyen (2001). Kou and Wang (2001) obtained results for the special case of the so-called double-exponential jump diffusion. Due to the memoryless property of the exponential distribution, they manage to obtained formulas for the Laplace transform of lookback and barrier options.

9.2 Other Exotic Options

Next, we list some results in the literature for some other exotic options. Most results are obtained making specific model restrictions.

9.2.1 The Perpetual American Call and Put Option

An American perpetual option is a contract between two parties, in which the first one, the holder, buys the right to receive from the other party, the seller, at a future time τ of his choosing an amount $G(S_\tau)$. Call and put options have the reward functions $G(x) = (x - K)^+$ and $G(x) = (K - x)^+$, respectively.

The optimal τ will depend on the evolution of the stock prices and as such is a random variable. In classical American options, the contract includes an exercise time T, the maturity, at or before the holder can exercise: $0 \leqslant \tau \leqslant T$. In the perpetual case $T = \infty$, so there is no expiry.

In Boyarchenko and Levendorskiĭ (2000, 2002a) some explicit formulas were derived using the theory of pseudo-differential operators.

Using probabilistic techniques, Mordecki (2002) studied perpetual American call and put options in terms of the overall supremum or infinimum of the Lévy process. Explicit formulas were obtained by Mordecki (2002) under the assumption of mixed-exponentially distributed and arbitrary negative jumps for the call options, and negative mixed-exponentially distributed and arbitrary positive jumps for put options.

These results generalize the closed formulas of McKean (1965) and Merton (1973) for the Black–Scholes setting.

9.2.2   The Perpetual Russian Option

The perpetual Russian option is an American-type option with no expiry offering the holder to exercise at any $\mathbb{F}$-stopping time, τ, in order to claim

$$\exp(-\alpha\tau)\max\left\{K, \sup_{0\leqslant u\leqslant \tau} S_u\right\}.$$

We say that it is an option with reduced regret.

The price of a Russian option in the traditional Black–Scholes market can be found in the original paper of Shepp and Shiryaev (1994). See also Shepp and Shiryaev (1993).

This option is studied for a Lévy market in Avram *et al.* (2003), Kou and Wang (2001) and Mordecki and Moreira (2002) for a jump diffusion with only negative jumps.

Asmussen *et al.* (2001) work with a class of Lévy processes which may have jumps in both directions but where the jumps are in the dense class of phase-type distributions (see Neuts 1981).

9.2.3   Touch-and-Out Options

A touch-and-out option (another name is first-touch digital) pays one currency unit the first time the stock price hits or crosses a predetermined level H from above. In other words, if the stock enters the zone $(0, H]$, the holder receives one currency unit. If the stock price always stays above H before expiry, the claim expires worthless.

Similar option contracts can be constructed which pay out the first time the stock price crosses the level H from below. Explicit pricing formulas under the Black–Scholes model can be found in, for example, Ingersoll (2000).

Option prices under a Lévy market were obtained by Boyarchenko and Levendorskiĭ (2002a). A detailed comparison of NIG-based prices with Black–Scholes prices can be found in Kudryavtsev and Levendorskiĭ (2002).

Boyarchenko and Levendorskiĭ (2002a) generalize the formulas for power first-touch contracts and contracts which pay a nonzero amount when a first barrier has been crossed but when a second one has not, and expire worthless if both barriers have been crossed in one jump.

9.3   Exotic Option Pricing by Monte Carlo Simulation

9.3.1   Introduction

If no closed formulas are at hand, we try to find a good indication of the price of the option by doing a huge number of simulations. The accuracy of the final estimate depends upon the number of sample paths used.

The method is basically as follows. Using the techniques described in the previous chapter we simulate, say m, paths of our stock-prices process and calculate for each path the value of the payoff function $V_i, i = 1, \ldots, m$. Then the Monte Carlo estimate of the expected value of the payoff is

$$\hat{V} = \frac{1}{m} \sum_{i=1}^{m} V_i. \tag{9.5}$$

The final option price is then obtained by discounting this estimate: $\exp(-rT)\hat{V}$.

The standard error of the estimate is given by

$$\sqrt{\frac{1}{(m-1)^2} \sum_{i=1}^{m} (\hat{V} - V_i)^2}.$$

The standard error decreases with the square root of the number of sample paths: to reduce the standard error by half, it is necessary to generate four times as many sample paths.

Next, we work out in detail the procedure for pricing a European exotic option with time to maturity T and payoff function $G(\{S_u, 0 \leqslant u \leqslant T\})$. We use the techniques described in the previous chapter to simulate paths of our stock-prices process.

9.3.2   Monte Carlo Pricing

Monte Carlo Pricing under the BNS Models

Under the BNS models (also called Black–Scholes SV models) we have the following procedure.

1. Calibrate the model on the available vanilla option prices in the market, i.e. find the risk-neutral parameters of the model which give in some sense (for example, the smallest RMSE) the best model prices compared with the market prices.

2. With the parameters of step 1, do the following.

   (a) Simulate a significant number m of paths of the stock-price process $S = \{S_t, 0 \leqslant t \leqslant T\}$ by simulating the log price process via a simulation of the OU process (see Section 8.4.9).

   (b) For each path i calculate the payoff function $g_i = G(\{S_u, 0 \leqslant u \leqslant T\})$.

3. Calculate by (9.5) the mean of the sample payoffs to get an estimate of the expected payoff:

$$\hat{g} = \frac{1}{m} \sum_{i=1}^{m} g_i.$$

4. Discount the estimated payoff at the risk-free rate to get an estimate of the value of the derivative: $\exp(-rT)\hat{g}$.

Monte Carlo Pricing under the Lévy SV Models

Under the Lévy SV models we can follow the same procedure. We refine step 2(a).

1. Calibrate the model on the available vanilla option prices in the market, i.e. find the risk-neutral parameters of the model which give in some sense (for example, the smallest RMSE) the best model prices compared with the market prices.

2. With the parameters of step 1, do the following.

   (a) Simulate a significant number m of paths of the stock-price process $S = \{S_t, 0 \leqslant t \leqslant T\}$ by simulating the log price process via a simulation of the time-changing process:

   (i) simulate the rate of time change process $y = \{y_t, 0 \leqslant t \leqslant T\}$;
   (ii) calculate from (i) the time change $Y = \{Y_t = \int_0^t y_s \, ds, 0 \leqslant t \leqslant T\}$;
   (iii) simulate the Lévy process $X = \{X_t, 0 \leqslant t \leqslant Y_T\}$ (note that we sample over the period $[0, Y_T]$);
   (iv) calculate the time-changed Lévy process X_{Y_t}, for $t \in [0, T]$;
   (v) calculate the stock-price process $S = \{S_t, 0 \leqslant t \leqslant T\}$.

   (b) For each path i calculate the payoff function $g_i = G(\{S_u, 0 \leqslant u \leqslant T\})$.

3. Calculate by (9.5) the mean of the sample payoffs to get an estimate of the expected payoff:

$$\hat{g} = \frac{1}{m} \sum_{i=1}^{m} g_i.$$

4. Discount the estimated payoff at the risk-free rate to get an estimate of the value of the derivative: $\exp(-rT)\hat{g}$.

A sample of all ingredients in the case of the Meixner–CIR combination is shown in Figure 9.1: the rate of time change $y = \{y_t, 0 \leqslant t \leqslant T\}$, the stochastic business time $Y = \{Y_t, 0 \leqslant t \leqslant T\}$, the Lévy process $X = \{X_t, 0 \leqslant t \leqslant Y_T\}$, the time-changed Lévy process $\{X_{Y_t}, 0 \leqslant t \leqslant T\}$, and finally the stock-price process $S = \{S_t, 0 \leqslant t \leqslant T\}$.

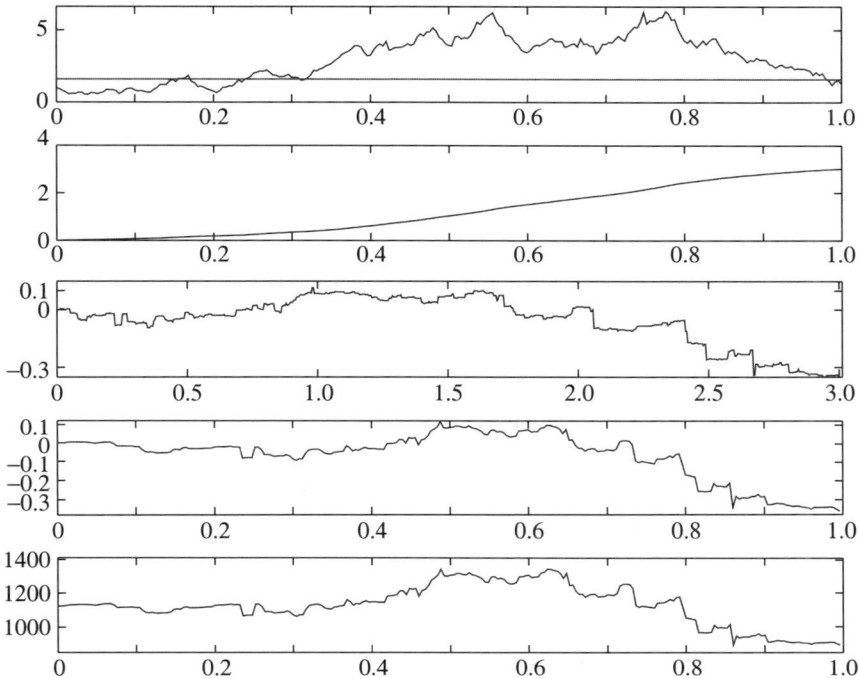

Figure 9.1 Simulation of y_t, Y_t, X_t, X_{Y_t} and S_t.

9.3.3 Variance Reduction by Control Variates

If we want to price exotic barrier and lookback options or other exotics (of European type), we often have information on vanilla options available. Note that we have obtained our parameters from calibration of market vanilla prices. In this case, where we thus have exact pricing information on related objects, we can use the variance reduction technique of control variates. The method speeds up the pricing, but the implementation depends on the characteristics of the instruments being valued.

The idea is as follows. Let us assume that we wish to calculate some expected value, $E[G] = E[G(\{S_t, 0 \leqslant t \leqslant T\})]$, of a (payoff) function G and that there is a related function H whose expectation $E[H] = E[H(\{S_t, 0 \leqslant t \leqslant T\})]$ we know exactly. Think of G as the payoff function of the exotic option we want to price via the Monte Carlo method and of H as the payoff function of the vanilla option whose price (and thus the expectation $E[H]$) we observe in the market.

Suppose that for a sample path the values of the functions G and H are positively correlated, e.g. the value of an up-and-in call is positively correlated with the value of a vanilla call with same strike price and time to expiry. This can be seen, for example, from Equation (9.1).

Define for some number $\beta \in \mathbb{R}$ a new payoff function

$$\hat{G}(\{S_t, 0 \leqslant t \leqslant T\}) = G(\{S_t, 0 \leqslant t \leqslant T\}) + \beta(H(\{S_t, 0 \leqslant t \leqslant T\}) - E[H]).$$

Note that the expected value of the new function $\hat{G}$ is the same as the expectation of the original function G. However, there can be a significant difference in the variance. We have

$$\text{var}[\hat{G}] = \text{var}[G] - 2\beta \, \text{cov}[G, H] + \beta^2 \, \text{var}[H].$$

This variance is minimized if

$$\beta = \frac{\text{cov}[G, H]}{\text{var}[H]} = \frac{E[GH] - E[G]E[H]}{\text{var}[H]}.$$

For this minimizing value of β we find

$$\text{var}[\hat{G}] = \text{var}[G]\left(1 - \frac{\text{cov}^2[G, H]}{\text{var}[G] \, \text{var}[H]}\right)$$

$$= \text{var}[G](1 - \text{corr}^2(G, H))$$

$$\leqslant \text{var}[G].$$

So, if the absolute value of the correlation between G and H is close to 1, the variance of $\hat{G}$ will be very small. Clearly, if we find such a highly correlated function H, very large computational savings may be made. H is called the control variate (CV). Note that the method is flexible enough to include several CVs.

The precise optimal value for β is not known but can be estimated from the same simulation. Special care has to be taken, however, since estimating parameters determining the result from the same simulation can introduce a bias. In the limit of a very large number of iterations, this bias vanishes. A remedy for the problem of bias due to the estimation of β is to use an initial simulation, possibly with fewer iterates than the main run, to estimate β in isolation. The CV technique usually provides such a substantial speed-up in convergence that this initial parameter estimation is affordable.

To summarize, we give an overview of the procedure (with an initial estimation of β). Recall that we want to price a European exotic option expiring at time T with payoff function $G(\{S_t, 0 \leqslant t \leqslant T\})$ and that we have a correlated option also expiring at time T with payoff $H(\{S_t, 0 \leqslant t \leqslant T\})$ whose option price is observable in the market and is given by

$$\exp(-rT)E[H(\{S_t, 0 \leqslant t \leqslant T\})] = \exp(-rT)E[H].$$

The expectation is under the risk-neutral pricing measure. We proceed as follows.

1. Estimate the optimal β:

   (a) sample a significant number n of paths for the stock price $S = \{S_t, 0 \leqslant t \leqslant T\}$ and calculate for each path i: $g_i = G(\{S_t, 0 \leqslant t \leqslant T\})$ and $h_i = H(\{S_t, 0 \leqslant t \leqslant T\})$;

Table 9.1 Exotic option prices.

Model	LC	UIB	UOB	DIB	DOB
VG–CIR–EWI	135.27	78.50	63.18	17.71	86.07
	(0.4942)	(0.2254)	(0.6833)	(0.5306)	(0.0811)
NIG–CIR–EWI	135.24	79.08	63.54	16.47	86.12
	(0.4764)	(0.2161)	(0.6665)	(0.4924)	(0.0819)
Meixner–CIR–EWI	135.72	78.57	64.34	17.28	86.06
	(0.4853)	(0.2239)	(0.7091)	(0.5168)	(0.0836)
VG–CIR–IILB	134.77	78.66	62.89	17.42	86.16
	(0.4894)	(0.2224)	(0.7250)	(0.5259)	(0.0958)
NIG–CIR–IILB	135.48	78.66	63.27	16.76	86.18
	(0.4817)	(0.2203)	(0.6841)	(0.5409)	(0.0609)
Meixner–CIR–IILB	134.83	78.66	63.87	17.24	86.08
	(0.4712)	(0.2193)	(0.7087)	(0.5560)	(0.0794)
Black–Scholes σ_{min}	128.64	65.66	46.26	21.78	69.17
Black–Scholes σ_{lse}	155.12	81.66	39.98	32.88	83.50
Black–Scholes σ_{max}	189.76	102.20	30.32	54.73	101.60

(b) an estimate for β is

$$\hat{\beta} = \frac{\sum_{i=1}^{n} g_i h_i - E[H] \sum_{i=1}^{n} g_i}{\sum_{i=1}^{n} (h_i - E[H])^2}.$$

2. Simulate a significant number m of paths for the stock price

$$S = \{S_t, 0 \leqslant t \leqslant T\}$$

and calculate for each path i:

$$g_i = G(\{S_t, 0 \leqslant t \leqslant T\}) \quad \text{and} \quad h_i = H(\{S_t, 0 \leqslant t \leqslant T\}).$$

3. Calculate an estimation of the expected payoff by

$$\hat{g} = \frac{1}{n} \left(\sum_{i=1}^{m} g_i - \hat{\beta}(h_i - E[H]) \right).$$

4. Discount the estimated payoff $\hat{g}$ at the risk-free rate r to get an estimate of the value of the derivative. The option price is given by $\exp(-rT)\hat{g}$.

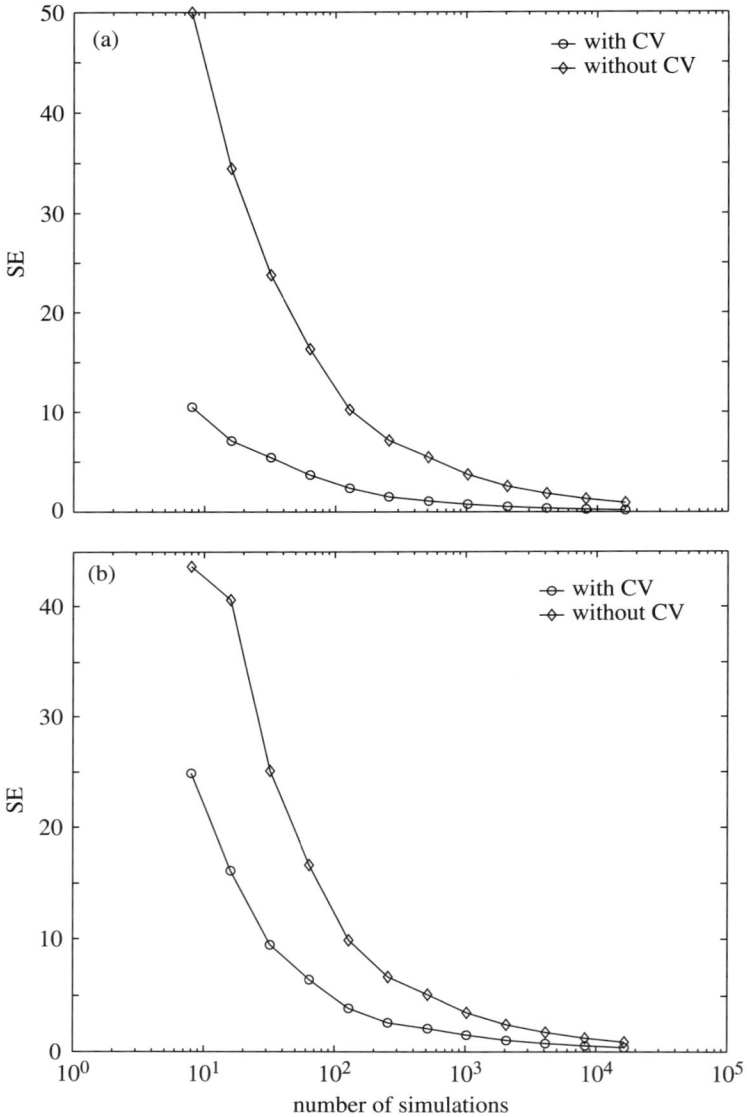

Figure 9.2 Standard error with and without CVs. (a) Up-and-in barrier; (b) lookback call.

9.3.4 Numerical Results

In this section we calculate some prices of exotic options by the above Monte Carlo method. We focus on the VG, the NIG and the Meixner cases in combination with the CIR process. The parameters of the processes are again as given in Table 7.3.

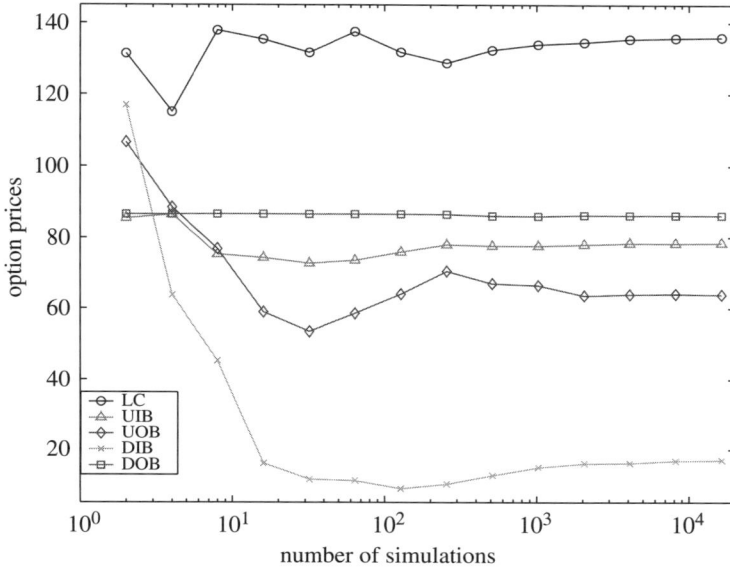

Figure 9.3 Convergence of options prices.

For all barrier options we take the time to maturity $T = 1$, the strike $K = S_0$ and the barrier H:

$$H_{UIB} = 1.1 \times S_0, \qquad H_{UOB} = 1.3 \times S_0,$$
$$H_{DIB} = 0.95 \times S_0, \qquad H_{DOB} = 0.8 \times S_0.$$

For all models, we make $n = 10\,000$ simulations of paths covering a one-year period. The time is discretized in 250 equally small time steps. We run 100 simulations to find an estimate for the optimal β of the control variate. We consider both equally weighted intervals (EWIs) and intervals with inverse linear boundaries (IILBs).

In Table 9.1 we compare the price along all models considered together with Black–Scholes prices. The standard error of the simulation is given in brackets below each option price. The volatility parameter in the Black–Scholes model is taken to be $\sigma_{lse} = 0.1812$, $\sigma_{min} = 0.1479$ and $\sigma_{max} = 0.2259$. These σs, which can be read off from Figure 4.5, correspond to the volatility giving rise to the least-square error of the Black–Scholes model prices compared with the empirical S&P 500 vanilla options, and the minimal and maximal implied volatility parameters over all strikes and maturities of our dataset, respectively. The Black–Scholes barrier prices are adjusted for the discrete observation of the stock prices as described above.

The effect of using control variates for the Monte Carlo pricing of the up-and-in barrier and the lookback option in the Meixner–CIR case is shown in Figure 9.2. Similar figures can be obtained for the other options and cases; all show that the standard error declines much faster with control variates than without. In Figure 9.3 we can see how the Monte Carlo prices converge over the number of iterations in the Meixner–CIR case. Note that in both figures we have logarithmic scales for the number of iterations.

9.3.5 Conclusion

First we note that all the SV models give very good fit to the data. If we look at the pricing of the exotic options in the Black–Scholes world, we observe that the Black–Scholes prices depend heavily on the choice of the volatility parameter and that it is not clear which value we should take. For the Lévy SV models the prices are very close to each other. We conclude that the Black–Scholes model is not at all appropriate for the pricing of exotics. Moreover, there is evidence that the Lévy SV models are much more reliable; they give a much better indication of the true price than the Black–Scholes model.

10

Interest-Rate Models

In this chapter we will look at models for describing the stochastic behaviour of interest rates. Typically, the classical models are often based on assumptions that returns in the bond market are (approximately) Normally distributed. We refer to Björk (1998), Brigo and Mercurio (2001), Filipović (2001), James and Webber (2000), Bingham and Kiesel (1998) and Rebonato (1996) for a comprehensive introduction to interest-rate modelling.

Empirical studies show that the normality assumptions of the classical models do not hold in general. We will describe the Lévy-based models introduced by Eberlein and Raible (1999) (see also Raible 2000).

We start with the general theory of interest-rate markets. Next, we give an overview of the classical Gaussian Heath–Jarrow–Morton (HJM) model and give some idea of its shortcomings. In order to deal with these drawbacks, we introduce models based on Lévy processes. Next, we indicate how to obtain prices of European vanilla options on bonds. Finally, we look at multi-factor extensions of the models presented.

10.1 General Interest-Rate Theory

Zero-Coupon Bonds

Denote by $P_t(T)$ the value at time t of one currency unit received for sure at time T. $P_t(T)$ is the value at time t of the so-called *zero-coupon bond* maturing with value 1 at time T. It is the discounting factor for cashflows occurring at time T. These bonds do not pay interest periodically, but give a face value (for convenience we have taken one currency unit) which will be paid at maturity; the interest earned on this bond appears as a discount to the face value at the beginning.

Interest rates are not a one-dimensional object. In the market there are bonds with maturities between 0 and 30 years (or even more). The interest received depends on the time to maturity. Under normal circumstances the interest rate paid for a bond with many years to maturity is higher than that for a bond which is close to maturity. We thus need to model interest rates with a vector- or function-valued stochastic process.

Lévy Processes in Finance   W. Schoutens
© 2003 John Wiley & Sons, Ltd   ISBN: 0-470-85156-2

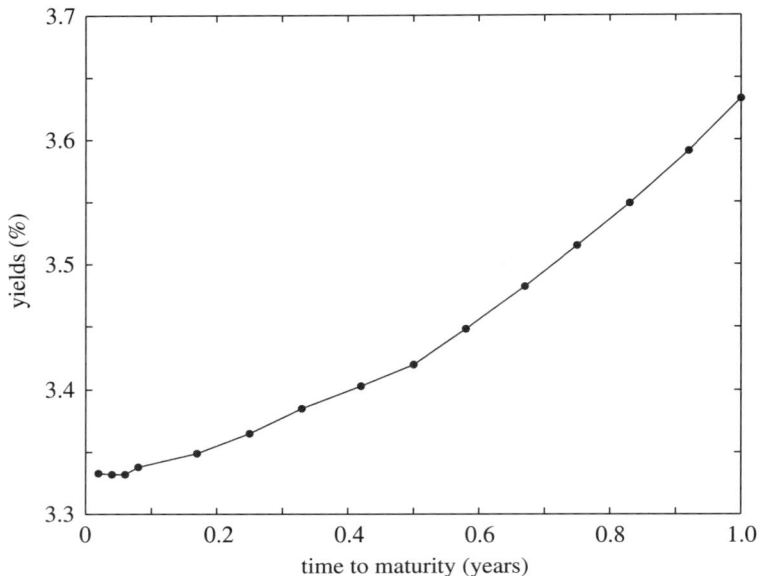

Figure 10.1   Yield curve on 26 February 2002.

We assume that there is a complete set of zero-coupon bonds with maturities T in the full time interval $[0, T^*]$, for some T^* (for example, 30 years). Given (at time t) a set of zero-coupon-bond prices $\{P_t(T), t < T \leqslant T^*\}$, the term structure of interest rates is the set of yields to maturity $\{r_t(T), t < T \leqslant T^*\}$ given (using continuous compounding) by

$$r_t(T) = -\frac{1}{T-t} \log P_t(T), \quad t < T \leqslant T^*.$$

This is known as the yield curve. In Figure 10.1, we see, for example, the EURI-BOR (Euro Interbank Offered Rate) yield curve on 26 February 2002 for bonds with maturity up to one year. EURIBOR is the rate at which Euro interbank term deposits within the Euro zone are offered by one prime bank to another prime bank.

The Short Rate

The short rate $r_t = r_t(t) = \lim_{T \downarrow t} r_t(T)$ is the rate on instantaneous borrowing and lending. Historically, it was the short rate which was modelled as the basic process. Although we have assumed in previous chapters that the short rate r is constant, in practice, this rate is stochastic and can fluctuate over time: $r = \{r_t, t \geqslant 0\}$. Note that the short rate is actually a theoretical entity which does not exist in real life and cannot be directly observed. Note also that in some markets the overnight interest rate is usually not considered a good approximant for the short rate. The driving motives and needs of overnight borrowing can be very different from the other rates.

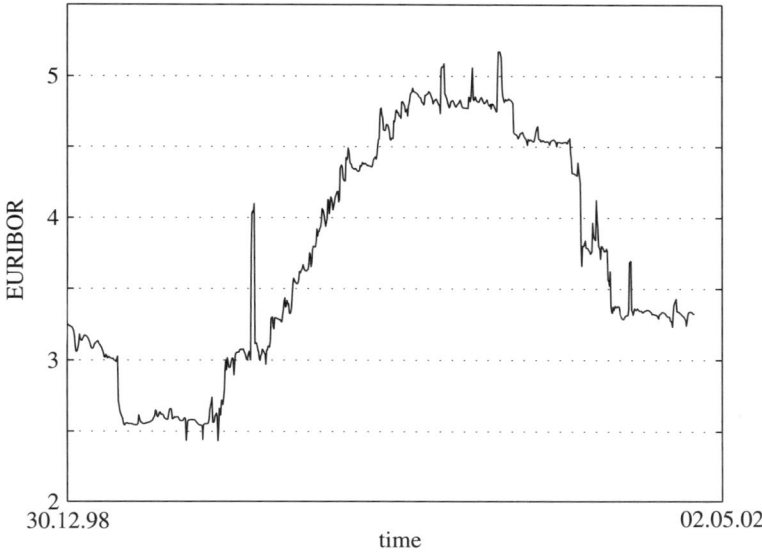

Figure 10.2   EURIBOR from 30 December 1998 until 2 May 2002.

In particular, its correlation with rates further up the term structure may be very slight. The one-week or one-month or even three-month interest rate is often a reasonable proxy.

The fact that interest rates behave stochastically can, for example, be seen from Figure 10.2, which shows the interest-rate fluctuations of the one-week EURIBOR interest rate from 30 December 1998 until 2 May 2002. The one-year USA Treasury yield over the period 1970–2001 is shown in Figure 10.3.

A sum of 1 invested in the short rate at time zero and continuously rolled over, i.e. instantaneously reinvested, is called the money-market account. Its value p_t at time t is

$$p_t = \exp\left(\int_0^t r_s \, ds\right).$$

If r is deterministic and constant, p_t reduces to our classical bank account: $p_t = B_t = \exp(rt)$.

A basic model for the behaviour of the stochastic process r is the CIR model. It is based on the CIR process, which we also encountered as a possible time change in Chapter 7:

$$dr_t = \kappa(\eta - r_t) \, dt + \sigma r_t^{1/2} \, dW_t, \quad r_0 > 0.$$

Under this model, the interest rate is mean reverting; it fluctuates around a long-term mean η. In the literature, this model was generalized in different ways. Well-known alternatives are the Vasicek and the Ho–Lee models, together with their extensions. We refer to Björk (1998).

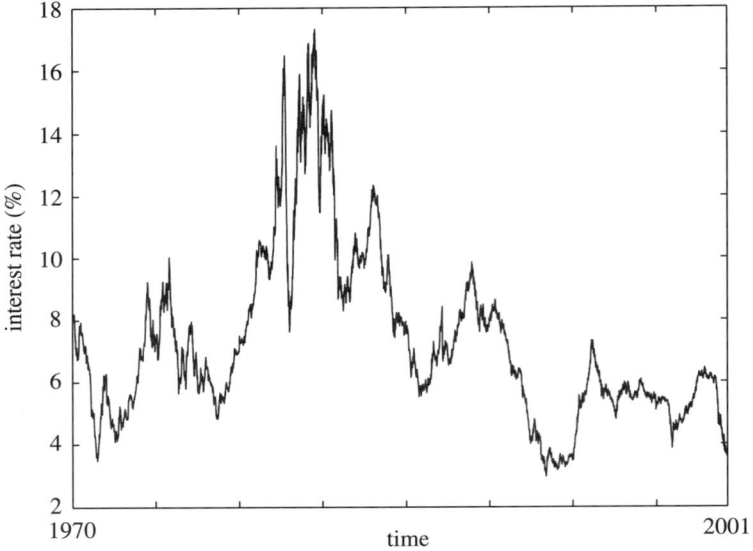

Figure 10.3 One-year USA Treasury yield from 1970 to 2001.

10.2   The Gaussian HJM Model

Instead of using the short rate as a state variable, Heath, Jarrow and Morton (HJM) proposed in Heath *et al.* (1992) to use the entire (forward) rate curve as the (infinite-dimensional) state variable. In the HJM model an entire rate curve evolves simultaneously. Moreover, the HJM model uses all the information available in the initial term structure.

The Instantaneous Forward Rate

We define the instantaneous forward rate to be

$$f(t, T) = -\frac{\partial}{\partial T} \log P_t(T).$$

The function $f(t, T)$ corresponds to the rate we can contract for at time t on a riskless loan that begins at time T and is returned an instant later. Since

$$P_t(T) = \exp\left(-\int_t^T f(t, s) \, ds\right),$$

zero-coupon-bond prices and forward rates represent equivalent information. Note that the short rate r_t is contained in this forward rate structure since $r_t = f(t, t)$.

The HJM Model

We start with the Heath, Jarrow and Morton (HJM) model for the forward rate. We assume that the forward rate dynamics are given by

$$df(t, T) = \alpha(t, T)\, dt + v(t, T)\, dW_t, \quad f(0, T) > 0, \tag{10.1}$$

where $\{W_t, t \geqslant 0\}$ is a standard Brownian motion and the functions α and v are sufficiently smooth. In general we can take an n-dimensional Brownian motion, taking into account various factors. Moreover, the drift function $\alpha(t, T)$ and volatility function $v(t, T)$ can be made path dependent. We focus first on the one-dimensional case with deterministic drift and volatility functions, and then, in Section 10.5, we consider the general case.

The above dynamics of the forward rates can be translated to the zero-coupon-bond prices, which are driven by

$$dP_t(T) = P_t(T)(m(t, T)\, dt + \sigma(t, T)\, dW_t), \quad P_0(T) > 0. \tag{10.2}$$

The relation between (10.1) and (10.2) is given by

$$m(t, T) = f(t, t) - \int_t^T \alpha(t, s)\, ds + \frac{1}{2}\left(\int_t^T v(t, s)\, ds\right)^2,$$

$$\sigma(t, T) = -\int_t^T v(t, s)\, ds.$$

From (10.1), we can recover the dynamics of the short rate under this model as

$$r_t = f(0, t) + \int_0^t \alpha(s, t)\, ds + \int_0^t v(s, t)\, dW_s, \quad r_0 > 0.$$

The HJM Drift Conditions

To rule out arbitrage, forwards drifts and volatilities (or, equivalently, bond price drifts and volatilities) must be consistent. This leads to the HJM drift conditions. Under a risk-neutral setting, we must have

$$\alpha(t, T) = v(t, T)\int_t^T v(t, s)\, ds, \quad 0 \leqslant t \leqslant T \leqslant T^*.$$

In terms of bond prices this means that the drift coefficient $m(t, T)$ is replaced by the short rate r_t. Therefore, in the risk-neutral world and in the HJM model the zero-coupon-bond prices satisfy an SDE of the form

$$dP_t(T) = P_t(T)(r_t\, dt + \sigma(t, T)\, dW_t), \quad P_0(T) > 0. \tag{10.3}$$

We assume that $P_0(T)$ as well as the (nonrandom) volatility $\sigma(s, T)$ are sufficiently smooth, namely at least C^2 and $\sigma(s, s) = 0$.

As in the Black–Scholes model (for stocks), the HJM model requires the underlying asset (the initial term structure) and a measure of its volatility as the only inputs.

The Volatility Structure

Note that contrary to the case of stock-price models it would not make sense to consider a constant volatility σ. When a default-free bond approaches its maturity, the span of possible price fluctuations narrows. This is clear since at maturity the owner of the bond will get the face value with certainty. Note that we have $\sigma(T, T) = 0$.

We will consider stationary volatility structures, i.e. $\sigma(t, T)$ depends only on the difference $T - t$. The Vasicek and Ho–Lee volatility structures are often used, and are given by, respectively,

$$\sigma(t, T) = \frac{b}{a}(1 - \exp(-a(T - t))),$$

$$\sigma(t, T) = b(T - t),$$

for parameters $b > 0$ and $a \neq 0$.

Bond Prices in the Gaussian HJM Model

The solution of the SDE (10.3) can be written in the form

$$P_t(T) = P_0(T) \exp\left(\int_0^t r_s \, ds \right) \frac{\exp(\int_0^t \sigma(s, T) \, dW_s)}{E[\exp(\int_0^t \sigma(s, T) \, dW_s)]} \tag{10.4}$$

$$= P_0(T) \exp\left(\int_0^t r_s \, ds + \int_0^t \sigma(s, T) \, dW_s - \int_0^t \tfrac{1}{2}\sigma(s, T)^2 \, ds \right) \tag{10.5}$$

$$= P_0(T) p_t \exp\left(\int_0^t \sigma(s, T) \, dW_s - \int_0^t \tfrac{1}{2}\sigma(s, T)^2 \, ds \right). \tag{10.6}$$

By Equation (10.5), the log return between times t and $t + \Delta t$ on a zero-coupon bond maturing at time T is given by

$$\log P_{t+\Delta t}(T) - \log P_t(T)$$
$$= \int_t^{t+\Delta t} r_s \, ds + \int_t^{t+\Delta t} \sigma(s, T) \, dW_s - \int_t^{t+\Delta t} \tfrac{1}{2}\sigma(s, T)^2 \, ds.$$

For $\Delta t \to 0$, the integrals may be replaced by the product of the value of the integrand at the left endpoint times the increment of the integrator. Hence, for Δt small enough, we have

$$\log P_{t+\Delta t}(T) - \log P_t(T) \approx r_t \Delta t + \sigma(t, T)(W_{t+\Delta t} - W_t) - \tfrac{1}{2}\sigma(t, T)^2 \Delta t.$$

We conclude that the log returns under the risk-neutral measure approximately follow a Normally distributed random variable:

$$\log P_{t+\Delta t}(T) - \log P_t(T) \sim \text{Normal}(r_t \Delta t - \tfrac{1}{2}\sigma(t, T)^2 \Delta t, \sigma(t, T)^2 \Delta t).$$

Raible (2000) argues that (under some assumptions and using Girsanov's theorem) that the objective (historical) measure will again approximately follow a Normal distribution.

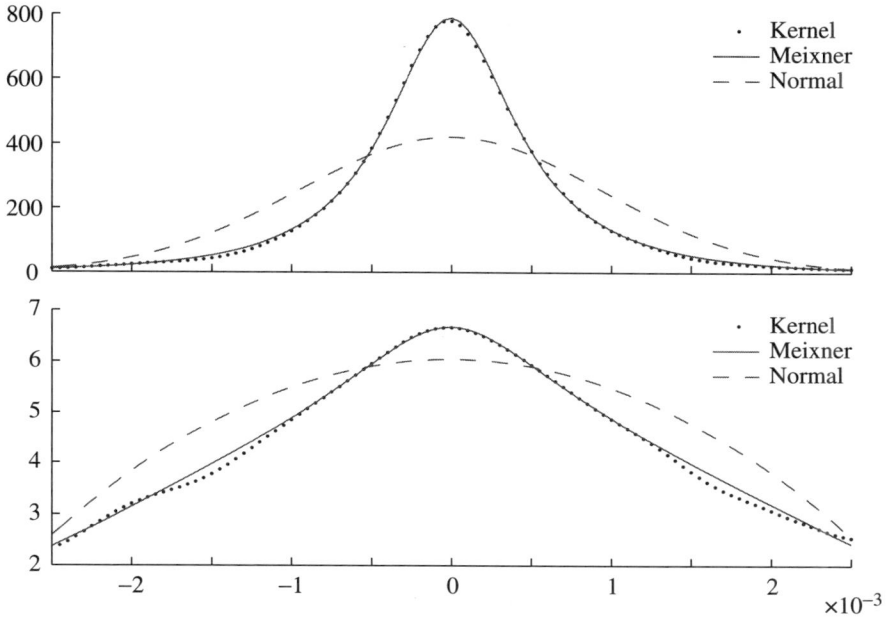

Figure 10.4 Density and log density for bond prices based on
the one-year USA Treasury yields from 1970 to 2001.

As in the case of stock prices, empirical studies (see, for example, Raible 2000) show that this normality assumption does not reflect reality. Empirically observed log returns of bonds turn out to have a leptokurtic distribution.

The density plots of the log returns of bond prices based on the one-year USA Treasury yields over the period 1970–2001 are shown in Figure 10.4. We see that the Gaussian kernel density and log density do not correspond at all to the Normal density, with mean and variances equal to the sample mean and sample variance. For a detailed study, including QQ plots and χ^2-tests, we refer to Raible (2000), who concludes that the Gaussian HJM model performs poorly as a description of the empirically observed movements of bond prices. However, the more flexible distributions already encountered in the modelling of stock-price log returns also give a much better fit here. This can, for example, also be seen from Figure 10.4, where a Meixner distribution has been fitted to the Gaussian kernel estimator. Clearly, the fit is significantly better than the Normal fit. More detailed studies (for the NIG case) can be found in Raible (2000).

10.3 The Lévy HJM Model

In order to obtain more realistic modelling, we replace the driving Brownian motion $W = \{W_t, t \geqslant 0\}$ in (10.4) with a more flexible Lévy process $L = \{L_t, t \geqslant 0\}$. In order to assume finiteness of the expectation in the denominator above in the case of general Lévy processes, we assume that

$$\int_{\{|x|>1\}} \exp(vx)\nu(dx) < \infty \quad \text{for } |v| < (1+\epsilon)M,$$

where $\epsilon > 0$ and M is such that $0 \leqslant \sigma(s, T) \leqslant M$ for $0 \leqslant s \leqslant T \leqslant T^*$ and $\nu(dx)$ is the Lévy measure of the infinitely divisible distribution of L_1. Typical choices of this Lévy process are the VG, the NIG, the GH, the Meixner or CGMY processes.

We now model the zero-coupon bond price with the following process:

$$P_t(T) = P_0(T) \exp\left(\int_0^t r_s \, ds\right) \frac{\exp(\int_0^t \sigma(s, T) \, dL_s)}{E[\exp(\int_0^t \sigma(s, T) \, dL_s)]}.$$

The Equivalent Martingale Measure

Using classical arguments from stochastic integration theory, we can easily prove that the discounted bond-price process,

$$\tilde{P}(T) = \left\{ \tilde{P}_t(T) = \exp\left(-\int_0^t r_s \, ds\right) P_t(T) = \frac{P_t(T)}{p_t}, \ 0 \leqslant t \leqslant T \right\},$$

is a martingale, and this in the general Lévy case, i.e. with W_t replaced by L_t.

Eberlein and Raible (1999) derived the bond-price process in the form,

$$P_t(T) = P_0(T) \exp\left(\int_0^t r_s \, ds\right) \frac{\exp(\int_0^t \sigma(s, T) \, dL_s)}{\exp(\int_0^t \theta(\sigma(s, T)) \, ds)},$$

where $\theta(u) = \log(E[\exp(uL_1)])$ denotes the logarithm of the moment-generating function of the Lévy process at time 1. In the classical Gaussian model we choose $\theta(u) = u^2/2$ and $L_s = W_s$.

As noted above, discounted bond prices are martingales in this term-structure model. It was shown in Raible (2000) that the martingale measure is unique. As a consequence, arbitrage-free prices of interest-rate derivatives are uniquely determined once the parameters of the driving Lévy process and the volatility structure are fixed.

10.4 Bond Option Pricing

In this section, we follow Eberlein and Raible (1999) and show how the price of vanilla European options can be obtained (numerically).

We concentrate on the time 0 price of a European call option on a bond maturing at time T, with exercise date t and strike price K. Using the classical arguments (and using the money-market account as numeraire), the price is given by

$$C(t, T, K) = E\left[\frac{1}{p_t}(P_t(T) - K)^+\right], \tag{10.7}$$

where the expectation is taken under the risk-neutral measure.

European Vanilla Call Option Price under the Gaussian HJM Model

In the Gaussian case, the option price (10.7) can be calculated explicitly (see, for example, Bingham and Kiesel 1998),

$$C(t, T, K) = P_0(T)N(b_2) - K P_0(t)N(b_2 - b_1),$$

where as usual N denotes the standard Normal cumulative distribution function as in (3.1), and where b_1 and b_2 are given by

$$b_1 = \left(\int_0^t (\sigma(s, T) - \sigma(s, t))^2 ds \right)^{1/2},$$

$$b_2 = \frac{\log P_0(T) - \log P_0(t) - \log K}{b_1} + \frac{b_1}{2}.$$

Calculating European Vanilla Call Option Price under the Lévy HJM Model

Next, we will look at the more general Lévy-driven models. First note that, since $P_t(t) = 1$, we have

$$\frac{1}{P_t} = P_0(t) \exp\left(-\int_0^t \theta(\sigma(s, t)) ds + \int_0^t \sigma(s, t) dL_s \right).$$

Thus, the expectation in (10.7) can be written as

$$E\left[\left(P_0(T) \exp\left(-\int_0^t \theta(\sigma(s, T)) ds + \int_0^t \sigma(s, T) dL_s \right) \right.\right.$$
$$\left.\left. - K P_0(t) \exp\left(-\int_0^t \theta(\sigma(s, t)) ds + \int_0^t \sigma(s, t) dL_s \right) \right)^+ \right]. \quad (10.8)$$

The only random variables appearing here are the two stochastic integrals:

$$X = \int_0^t \sigma(s, T) dL_s,$$

$$Y = \int_0^t \sigma(s, t) dL_s.$$

The remaining terms are deterministic and can be easily computed.

In order to evaluate the expectation (10.8), we need the information of the joint distribution of X and Y. Eberlein and Raible (1999) showed that the corresponding joint characteristic function is given by

$$E[\exp(iuX + ivY)] = \exp\left(\int_0^t \psi(u\sigma(s, T) + v\sigma(s, t)) ds \right),$$

where ψ is the cumulant characteristic function (or the characteristic exponent) of L_1, i.e. $\psi(u) = \log E[\exp(iuL_1)]$.

From this we can recover, using the inverse Fourier transform, the joint density of X and Y. The price of the vanilla European call option can then be obtained numerically. See Eberlein and Raible (1999) for a comparison (of the hyperbolic model) with the Gaussian case. They report that, typically, option prices as a function of the forward-price/strike ratio are W-shaped: at-the-money prices are lower, while in-the-money and out-of-the-money prices are higher than in the Gaussian model.

10.5   Multi-Factor Models

All of the above models are driven by a one-dimensional stochastic process. Moreover, the volatility structure in them is always deterministic and stationary.

In reality, the (log)returns of zero-coupon bonds of different maturities are not perfectly correlated. Multi-factor models can take this into account. The use of multiple factors and path-dependent volatility functions gives a certain flexibility since it can incorporate changes in the level, the slope and curvature of the term structure, though with each extra factor there is a considerable increase in complexity and practicality. In practice, we have to trade off precision and numerical tractability.

The Multi-Factor Gaussian HJM Model

In its most general form, the Gaussian HJM model specifies the instantaneous forward rate process as

$$\mathrm{d}f(t, T) = \alpha(t, T, \omega)\,\mathrm{d}t + \sum_{i=1}^{n} v_i(t, T, \omega)\,\mathrm{d}W_t^{(i)}, \quad f(0, T) > 0,$$

where $W^{(i)} = \{W_t^{(i)}, 0 \leqslant t\}$, $i = 1, \ldots, n$, are n independent standard Brownian motions. ω is a sample point in the sample space Ω. The functions $\alpha(t, T, \omega)$ and $v_i(t, T, \omega), i = 1, \ldots, n$, are path dependent; this is indicated by the ω in the notation. Since we have n sources of uncertainty, $W^{(i)}, i = 1, \ldots, n$, this is an n-factor model.

Bond prices then behave as

$$\mathrm{d}P_t(T) = P_t(T)\left(m(t, T, \omega)\,\mathrm{d}t + \sum_{i=1}^{n} \sigma_i(t, T, \omega)\,\mathrm{d}W_t^{(i)}\right), \quad P_0(T) > 0,$$

where

$$m(t, T, \omega) = f(t, t) - \int_t^T \alpha(t, s, \omega)\,\mathrm{d}s + \frac{1}{2}\sum_{i=1}^{n}\left(\int_t^T v_i(t, s, \omega)\,\mathrm{d}s\right)^2,$$

$$\sigma_i(t, T, \omega) = -\int_t^T v_i(t, s, \omega)\,\mathrm{d}s.$$

To rule out arbitrage we impose the HJM drift conditions:

$$\alpha(t, T, \omega) = \sum_{i=1}^{n} v_i(t, T, \omega) \int_{t}^{T} v_i(t, s, \omega) \, ds, \quad 0 \leqslant t \leqslant T \leqslant T^*.$$

In terms of bond prices this again comes down to replacing the drift coefficient $m(t, T, \omega)$ by the short rate r_t. In the risk-neutral setting, the zero-coupon-bond price dynamics are

$$dP_t(T) = P_t(T) \left(r_t \, dt + \sum_{i=1}^{n} \sigma_i(t, T, \omega) \, dW_t^{(i)} \right), \quad P_0(T) > 0. \tag{10.9}$$

In contrast with the models of the preceding sections, bond prices are not necessarily Markovian; there is a possible dependence on the past via ω.

The Multi-Factor Lévy HJM Model

Raible (2000) uses the same idea as above to present a multi-factor stochastic volatility Lévy-driven term-structure model. He starts not with the driving SDE, but again with the explicit bond-price formula of the Gaussian counterpart, in which he replaces the standard Brownian motions with a Lévy process.

So, assume that we have n independent Lévy processes $L^{(i)} = \{L_t^{(i)}, t \geqslant 0\}$, $i = 1, \ldots, n$, satisfying some regularity conditions similar to the above. The derived bond-price process (in the risk-neutral setting) is of the form

$$P_t(T) = P_0(T) p_t \frac{\exp(\sum_{i=1}^{n} \int_0^t \sigma_i(s, T, \omega) \, dL_s^{(i)})}{\exp(\sum_{i=1}^{n} \int_0^t \theta_i(\sigma_i(s, T)) \, ds)},$$

where $\theta_i(u) = \log \vartheta_i(u) = \log(E[\exp(u L_1^{(i)})])$ denotes the logarithm of the moment-generating function of the ith Lévy process at time 1. In the classical Gaussian model we choose $\theta_i(u) = u^2/2$ and $L_s^{(i)} = W_s^{(i)}$, $i = 1, \ldots, n$.

By construction, we also have here that the discounted bond prices

$$\tilde{P}(T) = \{P_t(T)/p_t, 0 \leqslant t \leqslant T\}$$

are martingales.

In this way, it is possible to capture the three key features of the empirical behaviour of the term structure: non-Normal return behaviour, multi-factor movement, and stochastic volatility.

Appendix A

Special Functions

A.1 Bessel Functions

A standard reference for Bessel functions is Abramowitz and Stegun (1968). Sometimes the function N_v, which is defined below, is also denoted by Y_v.

Bessel functions of the first kind $J_{\pm v}(z)$, of the second kind $N_v(z)$, and of the third kind $H_v^{(1)}(z)$ and $H_v^{(2)}(z)$ are solutions to the differential equation:

$$z^2 \frac{d^2 w}{dz^2} + z \frac{dw}{dz} + (z^2 - v^2)w = 0.$$

The function $J_v(z)$ can be written as the following series:

$$J_v(z) = (z/2)^v \sum_{k=0}^{\infty} \frac{(-z^2/4)^k}{k!\,\Gamma(v+k+1)},$$

and $N_v(z)$ satisfies

$$N_v(z) = \frac{J_v(z)\cos(v\pi) - J_{-v}(z)}{\sin(v\pi)},$$

where the right-hand side of this equation is replaced by its limiting value if v is an integer or zero. We also have

$$H_v^{(1)}(z) = J_v(z) + iN_v(z),$$
$$H_v^{(2)}(z) = J_v(z) - iN_v(z).$$

Lévy Processes in Finance   W. Schoutens
© 2003 John Wiley & Sons, Ltd   ISBN: 0-470-85156-2

Next, we list some useful properties:

$$J_{1/2}(z) = \sqrt{\frac{2}{\pi z}} \sin z,$$

$$J_{-1/2}(z) = \sqrt{\frac{2}{\pi z}} \cos z,$$

$$J_{3/2}(z) = \sqrt{\frac{2}{\pi z}} \left(\frac{\sin z}{z} - \cos z \right),$$

$$J_{-3/2}(z) = \sqrt{\frac{2}{\pi z}} \left(\sin z + \frac{\cos z}{z} \right),$$

$$J_{n+1/2}(z) = (-1)^n N_{-n-1/2}(z), \qquad n = 0, 1, 2, \ldots,$$

$$J_{-n-1/2}(z) = (-1)^{n-1} N_{n+1/2}(z), \qquad n = 0, 1, 2, \ldots.$$

A.2 Modified Bessel Functions

The modified Bessel functions of the first kind $I_{\pm v}(z)$ and of the third kind (also called MacDonald functions) $K_v(z)$ are solutions to the differential equation

$$z^2 \frac{d^2 w}{dz^2} + z \frac{dw}{dz} - (z^2 + v^2) w = 0.$$

The function $I_v(z)$ can be written as the following series,

$$I_v(z) = (z/2)^v \sum_{k=0}^{\infty} \frac{(z^2/4)^k}{k! \Gamma(v + k + 1)},$$

and $K_v(z)$ satisfies

$$K_v(z) = \frac{\pi}{2} \frac{I_v(z) - I_{-v}(z)}{\sin(v\pi)},$$

where the right-hand side of this equation is replaced by its limiting value if v is an integer or zero.

The Bessel function K_v can also be written in integral form as

$$K_v(z) = \frac{1}{2} \int_0^{\infty} u^{v-1} \exp(-\tfrac{1}{2} z(u + u^{-1})) \, du, \quad x > 0.$$

Next, we list some useful properties:

$$K_v(z) = K_{-v}(z),$$

$$K_{v+1}(z) = \frac{2v}{z} K_v(z) + K_{v-1}(z),$$

$$K_{1/2}(z) = \sqrt{\pi/2} \, z^{-1/2} \exp(-z),$$

$$K_v'(z) = -\frac{v}{z} K_v(z) - K_{v-1}(z).$$

A.3   The Generalized Hypergeometric Series

The *generalized hypergeometric series* $_pF_q$ is defined by

$$_pF_q(a_1,\ldots,a_p; b_1,\ldots,b_q; z) = \sum_{j=0}^{\infty} \frac{(a_1)_j \cdots (a_p)_j}{(b_1)_j \cdots (b_q)_j} \frac{z^j}{j!},$$

where $b_i \neq 0, -1, -2, \ldots, i = 1, \ldots, q$, and where

$$(a)_n = \frac{\Gamma(a+n)}{\Gamma(a)} = \begin{cases} 1, & \text{if } n = 0, \\ a(a+1) \cdots (a+n-1), & \text{if } n = 1, 2, 3, \ldots. \end{cases}$$

There are p numerator parameters and q denominator parameters. Clearly, the orderings of the numerator parameters and of the denominator parameters are immaterial.

A.4   Orthogonal Polynomials

Here we summarize the ingredients of some orthogonal polynomial $y_n(x)$ of degree n. We have the following orthogonality relations,

$$\int_S y_n(x) y_m(x) \rho(x)\, dx = d_n^2 \delta_{nm},$$

where S is the support of $\rho(x)$.

The constant a_n is the leading coefficient of $y_n(x)$. We denote the monic polynomial (i.e. with leading coefficient 1) by $\tilde{y}_n(x) = a_n^{-1} y_n(x)$.

A.4.1   Hermite polynomials with parameter

Notation	$H_n(x; t)$
Restrictions	$t > 0$
HF*	$H_n(x; t) = (\sqrt{2/t}x)^n\ _2F_0(-n/2, -(n-1)/2; ; -2t/x^2)$
GF**	$\sum_{n=0}^{\infty} H_n(x; t)\dfrac{z^n}{n!} = \exp(\sqrt{2}xz/\sqrt{t} - z^2)$
$\rho(x)$	$\exp(-x^2/(2t))/\sqrt{2\pi t}$
Support	$(-\infty, +\infty)$
d_n^2	$2^n n!$
a_n	$(\sqrt{2/t})^n$

*HF, Hypergeometric Function; **GF, Generating Function

A.4.2   Meixner–Pollaczek Polynomials

Notation	$P_n(x; \delta, \zeta)$

Restrictions	$\delta > 0, 0 < \zeta < \pi$
HF*	$P_n(x; \delta, \zeta) = \dfrac{(2\delta)_n}{n!} \exp(in\zeta) \, {}_2F_1(-n, \delta + ix; 2\zeta; 1 - \exp(-2i\zeta))$
GF**	$\displaystyle\sum_{n=0}^{\infty} P_n(x; \delta, \zeta) z^n = (1 - \exp(i\zeta)z)^{-\delta+ix}(1 - \exp(-i\zeta)z)^{-\delta-ix}$
$\rho(x)$	$\dfrac{(2\sin(\zeta))^{2\delta}}{2\pi} \exp((2\zeta - \pi)x)|\Gamma(\delta + ix)|^2$
Support	$(-\infty, +\infty)$
d_n^2	$\Gamma(n + 2\delta)/n!$
a_n	$(-2\sin(x))^n$

*HF, Hypergeometric Function; **GF, Generating Function

Appendix B

Lévy Processes

B.1   Characteristic Functions

Recall that a Lévy process $X = \{X_t, t \geqslant 0\}$ is completely defined by the infinitely divisible law of X_1.

Next, we give the characteristic functions for these infinitely divisible distributions. We consider distributions on the nonnegative integers, the positive half-line and on the real line. Note that, for the latter, an extra m parameter can be added (except for the Normal distribution) as described in Section 5.4. The characteristic function of the extended distribution is just the product of $\exp(ium)$ and the original characteristic function.

B.1.1   Distributions on the Nonnegative Integers

Distribution	$\phi(u) = E[\exp(iuX_1)]$
Poisson(λ)	$\exp(\lambda(\exp(iu) - 1))$

B.1.2   Distributions on the Positive Half-Line

Distribution	$\phi(u) = E[\exp(iuX_1)]$
Gamma(a, b)	$(1 - iu/b)^{-a}$
Exp(λ)	$(1 - iu/\lambda)^{-1}$
IG(a, b)	$\exp(-a(\sqrt{-2iu + b^2} - b))$
GIG(λ, a, b)	$K_\lambda^{-1}(ab)(1 - 2iu/b^2)^{\lambda/2}K_\lambda(ab\sqrt{1 - 2iu/b^2})$
TS(κ, a, b)	$\exp(ab - a(b^{1/\kappa} - 2iu)^\kappa)$

Lévy Processes in Finance   W. Schoutens
© 2003 John Wiley & Sons, Ltd   ISBN: 0-470-85156-2

B.1.3   Distributions on the Real Line

Distribution	$\phi(u) = E[\exp(iuX_1)]$
Normal(μ, σ^2)	$\exp(iu\mu)\exp(-\sigma^2 u^2/2)$
VG(σ, ν, θ)	$(1 - iu\theta\nu + \sigma^2\nu u^2/2)^{-1/\nu}$
VG(C, G, M)	$(GM/(GM + (M - G)iu + u^2))^C$
NIG(α, β, δ)	$\exp(-\delta(\sqrt{\alpha^2 - (\beta + iu)^2} - \sqrt{\alpha^2 - \beta^2}))$
CGMY(C, G, M, Y)	$\exp(C\Gamma(-Y)((M - iu)^Y - M^Y + (G + iu)^Y - G^Y))$
Meixner(α, β, δ)	$(\cos(\beta/2)/\cosh((\alpha u - i\beta)/2))^{2\delta}$
GZ$(\alpha, \beta_1, \beta_2, \delta)$	$\left(\dfrac{B(\beta_1 + i\alpha u/2\pi, \beta_2 - i\alpha u/2\pi)}{B(\beta_1, \beta_2)}\right)^{2\delta}$
HYP(α, β, δ)	$\left(\dfrac{\alpha^2 - \beta^2}{\alpha^2 - (\beta + iu)^2}\right)^{1/2}\dfrac{\mathrm{K}_1(\delta\sqrt{\alpha^2 - (\beta + iu)^2})}{\mathrm{K}_1(\delta\sqrt{\alpha^2 - \beta^2})}$
GH$(\lambda, \alpha, \beta, \delta)$	$\left(\dfrac{\alpha^2 - \beta^2}{\alpha^2 - (\beta + iu)^2}\right)^{\lambda/2}\dfrac{\mathrm{K}_\lambda(\delta\sqrt{\alpha^2 - (\beta - iu)^2})}{\mathrm{K}_\lambda(\delta\sqrt{\alpha^2 - \beta^2})}$

B.2 Lévy Triplets

B.2.1 γ

Recall the Lévy–Khintchine formula for the logarithm of the characteristic function of X_1,

$$\log \phi(u) = i\gamma u - \tfrac{1}{2}\sigma^2 u^2 + \int_{-\infty}^{+\infty} (\exp(iux) - 1 - iux 1_{\{|x|<1\}})\nu(dx).$$

Next, we give the corresponding Lévy triplets $[\gamma, \sigma^2, \nu(dx)]$. $\sigma^2 = 0$ for all examples mentioned below, except for the Normal(μ, σ^2) distribution, where it is equal to the variance σ^2.

Distribution	γ
Normal(μ, σ^2)	0
Poisson(λ)	0
Gamma(a, b)	$a(1 - \exp(-b))/b$
IG(a, b)	$(a/b)(2N(b) - 1)$
GIG(λ, a, b)	$\displaystyle\int_0^1 \exp(-\tfrac{1}{2}b^2 x)$ $\displaystyle \times \left(\int_0^\infty \frac{\exp(-xz)}{\pi^2 z(J_{|\lambda|}^2(a\sqrt{2z}) + N_{|\lambda|}^2(a\sqrt{2z}))}\, dz + \max\{0, \lambda\} \right) dx$
TS(κ, a, b)	$\displaystyle a2^\kappa \frac{\kappa}{\Gamma(1-\kappa)} \int_0^1 x^{-\kappa} \exp(-\tfrac{1}{2}b^{1/\kappa}x)\, dx$
VG(C, G, M)	$C(MG)^{-1}(G(\exp(-M) - 1) - M(\exp(-G) - 1))$
NIG(α, β, δ)	$\displaystyle \frac{2\delta\alpha}{\pi} \int_0^1 \sinh(\beta x)K_1(\alpha x)\, dx$
CGMY(C, G, M, Y)	$\displaystyle C\left(\int_0^1 (\exp(-Mx) - \exp(-Gx))x^{-Y}\, dx \right)$
Meixner(α, β, δ)	$\displaystyle \alpha\delta \tan(\beta/2) - 2\delta \int_1^\infty \frac{\sinh(\beta x/\alpha)}{\sinh(\pi x/\alpha)}\, dx$

B.2.2 The Lévy Measure $\nu(dx)$

Distribution	$\nu(dx)$
Normal(μ, σ^2)	0
Poisson(λ)	$\lambda \delta(1)$
Gamma(a, b)	$a \exp(-bx) x^{-1} 1_{(x>0)} \, dx$
IG(a, b)	$(2\pi)^{-1/2} a x^{-3/2} \exp(-\frac{1}{2} b^2 x) 1_{(x>0)} \, dx$

GIG(λ, a, b)

$$x^{-1} \exp(-\tfrac{1}{2} b^2 x)$$

$$\times \left(\int_0^\infty \frac{\exp(-xz)}{\pi^2 z (J_{|\lambda|}^2(a\sqrt{2z}) + N_{|\lambda|}^2(a\sqrt{2z}))} \, dz + \max\{0, \lambda\} \right) 1_{(x>0)} \, dx$$

TS(κ, a, b)	$a 2^\kappa \dfrac{\kappa}{\Gamma(1-\kappa)} x^{-\kappa-1} \exp(-\frac{1}{2} b^{1/\kappa} x) 1_{(x>0)} \, dx$
VG(C, G, M)	$C|x|^{-1} (\exp(Gx) 1_{(x<0)} + \exp(-Mx) 1_{(x>0)}) \, dx$
NIG(α, β, δ)	$\delta \alpha \pi^{-1} |x|^{-1} \exp(\beta x) K_1(\alpha|x|) \, dx$
CGMY(C, G, M, Y)	$C|x|^{-1-Y} (\exp(Gx) 1_{(x<0)} + \exp(-Mx) 1_{(x>0)}) \, dx$
Meixner(α, β, δ)	$\delta x^{-1} \exp(\beta x/\alpha) \sinh^{-1}(\pi x/\alpha) \, dx$

GH$(\lambda, \alpha, \beta, \delta)$,
$\quad \lambda \geqslant 0$

$$\frac{\exp(\beta x)}{|x|} \left(\int_0^\infty \frac{\exp(-|x|\sqrt{2y + \alpha^2})}{\pi^2 y (J_\lambda^2(\delta\sqrt{2y}) + N_\lambda^2(\delta\sqrt{2y}))} \, dy + \lambda \exp(-\alpha|x|) \right)$$

GH$(\lambda, \alpha, \beta, \delta)$,
$\quad \lambda < 0$

$$\frac{\exp(\beta x)}{|x|} \int_0^\infty \frac{\exp(-|x|\sqrt{2y + \alpha^2})}{\pi^2 y (J_{-\lambda}^2(\delta\sqrt{2y}) + N_{-\lambda}^2(\delta\sqrt{2y}))} \, dy$$

Appendix C

S&P 500 Call Option Prices

In the following table, we can find 77 call option prices on the S&P 500 Index at the close of the market on 18 April 2002. On that day, the S&P 500 Index closed at 1124.47. We had values of $r = 1.9\%$ and $q = 1.2\%$ per year.

Strike	May 2002	June 2002	Sep. 2002	Dec. 2002	March 2003	June 2003	Dec. 2003
975			161.60	173.30			
995			144.80	157.00		182.10	
1025			120.10	133.10	146.50		
1050		84.50	100.70	114.80		143.00	171.40
1075		64.30	82.50	97.60			
1090	43.10						
1100	35.60		65.50	81.20	96.20	111.30	140.40
1110		39.50					
1120	22.90	33.50					
1125	20.20	30.70	51.00	66.90	81.70	97.00	
1130		28.00					
1135		25.60	45.50				
1140	13.30	23.20		58.90			
1150		19.10	38.10	53.90	68.30	83.30	112.80
1160		15.30					
1170		12.10					
1175		10.90	27.70	42.50	56.60		99.80
1200			19.60	33.00	46.10	60.90	
1225			13.20	24.90	36.90	49.80	
1250				18.30	29.30	41.20	66.90
1275				13.20	22.50		

Lévy Processes in Finance   W. Schoutens
© 2003 John Wiley & Sons, Ltd   ISBN: 0-470-85156-2

Strike	May 2002	June 2002	Sep. 2002	Dec. 2002	March 2003	June 2003	Dec. 2003
1300					17.20	27.10	49.50
1325					12.80		
1350						17.10	35.70
1400						10.10	25.20
1450							17.00
1500							12.20

References

Abramowitz, M. and Stegun, I. A. (1968) *Handbook of Mathematical Functions*. Dover, New York.

Applebaum, D. (2003) *Lévy Processes and Stochastic Calculus*. (In preparation.)

Artzner, P., Delbaen, F., Eber, J.-M. and Heath, D. (1999) Coherent measures of risk. *Mathematical Finance* **9**, 203–228.

Artzner, P. and Heath, D. (1995) Approximate completeness with multiple martingale measures. *Mathematical Finance* **5**, 1–11.

Asmussen, S., Avram, F. and Pistorius, M. R. (2001) *Russian Options under Exponential Phase-Type Lévy Models*. Working Paper.

Asmussen, S. and Rosiński, J. (2001) Approximations of small jumps of Lévy processes with a view towards simulation. *Journal of Applied Probability* **38**, 482–493.

Avram, F., Kyprianou, A. E. and Pistorius, M. R. (2003) Exit problems for spectrally negative Lévy processes and applications to (Canadized) Russian options. *Annals of Applied Probability*. (In the press.)

Bachelier, L. (1900) Théorie de la spéculation. *Annales Scientifiques l'École Normale Supérieure* **17**, 21–86.

Bar-Lev, S., Bshouty, D. and Letac, G. (1992) Natural exponential families and self-decomposability. *Statistics Probability Letters* **13**, 147–152.

Barndorff-Nielsen, O. E. (1977) Exponentially decreasing distributions for the logarithm of particle size. *Proceedings of the Royal Society of London* A **353**, 401–419.

Barndorff-Nielsen, O. E. (1978) Hyperbolic distributions and distributions on hyperbolae. *Scandinavian Journal of Statistics* **5**, 151–157.

Barndorff-Nielsen, O. E. (1995) Normal inverse Gaussian distributions and the modeling of stock returns. Research Report no. 300, Department of Theoretical Statistics, Aarhus University.

Barndorff-Nielsen, O. E. (1997) Normal inverse Gaussian distributions and stochastic volatility models. *Scandinavian Journal of Statistics* **24**, 1–13.

Barndorff-Nielsen, O. E. (1998) Processes of normal inverse Gaussian type. *Finance and Stochastics* **2**, 41–68.

Barndorff-Nielsen, O. E. (2001) Superposition of Ornstein–Uhlenbeck type processes. *Theory of Probability and Its Applications* **45**, 175–194.

Barndorff-Nielsen, O. E. and Halgreen, C. (1977) Infinitely divisibility of the hyperbolic and generalized inverse Gaussian distributions. *Zeitschrift für Wahrscheinlichkeitstheorie und verwandte Gebiete* **38**, 309–311.

Barndorff-Nielsen, O. E. and Shephard, N. (2001a) Non-Gaussian Ornstein–Uhlenbeck-based models and some of their uses in financial economics. *Journal of the Royal Statistical Society* B **63**, 167–241.

Barndorff-Nielsen, O. E. and Shephard, N. (2001b) Modelling by Lévy processes for financial econometrics. In *Lévy Processes – Theory and Applications* (ed. O. E. Barndorff-Nielsen, T. Mikosch and S. Resnick), pp. 283–318. Birkhäuser, Boston.

Barndorff-Nielsen, O. E. and Shephard, N. (2003a) Normal modified stable processes. *Theory of Probability and Mathematical Statistics*. (In the press.)

Barndorff-Nielsen, O. E. and Shephard, N. (2003b) Integrated OU Processes and non-Gaussian OU-based Stochastic Volatility Models. *Scandinavian Journal of Statistics*. (In the press.)

Barndorff-Nielsen, O. E., Mikosch, T. and Resnick, S. (eds) (2001) *Lévy Processes – Theory and Applications*. Birkhäuser, Boston.

Barndorff-Nielsen, O. E., Nicolato, E. and Shephard, N. (2002) Some recent developments in stochastic volatility modelling. *Quantitative Finance* **2**, 11–23.

Bakshi, G. and Madan, D. B. (2000) Spanning and derivative security valuation. *Financial Economics* **55**, 205–238.

Benth, F. E., Di Nunno, G., Løkka, A., Øksendal, B. and Proske, F. (2003) Explicit representation of minimal variance portfolio in markets driven by Lévy processes. *Mathematical Finance* **13**, 17–35.

Berman, M. B. (1971) Generating Gamma distributed variates for computer simulation models. Technical Report R-641-PR, Rand Corporation.

Bertoin, J. (1996) *Lévy Processes*. Cambridge Tracts in Mathematics, vol. 121. Cambridge University Press.

Biane, P., Pitman J. and Yor, M. (2001) Probability laws related to the Jacobi theta and Riemann zeta functions, and Brownian excursions. *Bulletin of the American Mathematical Society* **38**, 435–465.

Billingsley, P. (1995) *Probability and Measure*, 3rd edn. John Wiley & Sons, Ltd.

Bingham, N. H. (1998) Fluctuations. *Mathematical Scientist* **23**, 63–73.

Bingham, N. H. and Kiesel, R. (1998) *Risk-Neutral Valuation. Pricing and Hedging of Financial Derivatives*. Springer Finance, London.

Bingham, N. H. and Kiesel, R. (2001a) Modelling asset returns with hyperbolic distributions. In *Return Distributions on Finance* (ed. J. Knight and S. Satchell), pp. 1–20. Butterworth-Heinemann.

Bingham, N. H. and Kiesel, R. (2001b) Hyperbolic and semi-parametric models in finance. In *Disordered and Complex Systems*. American Institute of Physics.

Bingham, N. H. and Kiesel, R. (2002) Semi-parametric modelling in finance: theoretical foundations. *Quantitative Finance* **2**, 241–250.

Björk, T. (1998) *Arbitrage Theory in Continuous Time*. Oxford University Press.

Black, F. and Scholes, M. (1973) The pricing of options and corporate liabilities. *Journal of Political Economy* **81**, 637–654.

Blæsild, P. (1978) The shape of the generalized inverse Gaussian and hyperbolic distributions. Research Report no. 37, Department of Theoretical Statistics, Aarhus University.

Bouchaud, J.-P. and Potters, M. (1997) *Theory of Financial Risk*. Aléa-Saclay, Eurolles, Paris.

Boyarchenko, S. I. and Levendorskiĭ, S. Z. (1999) *Generalizations of the Black–Scholes equation for truncated Lévy processes*. Working paper.

Boyarchenko, S. I. and Levendorskiĭ, S. Z. (2000) Option pricing for truncated Lévy processes. *International Journal for Theory and Applications in Finance* **3**, 549–552.

Boyarchenko, S. I. and Levendorskiĭ, S. Z. (2002a) Perpetual American options under Lévy processes. *SIAM Journal of Control and Optimization* **40**, 1663–1696.

Boyarchenko, S. I. and Levendorskiĭ, S. Z. (2002b) *Non-Gaussian Merton–Black–Scholes Theory*. World Scientific.

Boyarchenko, S. I. and Levendorskiĭ, S. Z. (2002c) Barrier options and touch-and-out options under regular Lévy processes of exponential type. *Annals of Applied Probability* **12**, 1261–1298.

Brigo, D. and Mercurio, F. (2001) *Interest Rate Models: Theory and Practice*. Springer.

Broadie, M., Glasserman, P. and Kou, S. G. (1997) A continuity correction for discrete barrier options. *Mathematical Finance* **7**, 325–349.

Broadie, M., Glasserman, P. and Kou, S. G. (1999) Connecting discrete and continuous path-dependent options. *Finance and Stochastics* **3**, 55–82.

Bühlmann, H., Delbaen, F. Embrechts, P. and Shiryaev, A. N. (1996) No-arbitrage, change of measure and conditional Esscher transforms. *CWI Quarterly* **9**(4), 291–317.

Carr, P. and Madan, D. (1998) Option valuation using the fast Fourier transform. *Journal of Computational Finance* **2**, 61–73.

Carr, P., Geman, H., Madan, D. H. and Yor, M. (2002) The fine structure of asset returns: an empirical investigation. *Journal of Business* **75**, 305–332.

Carr, P., Geman, H., Madan, D. H. and Yor, M. (2003) Stochastic volatility for Lévy processes. *Mathematical Finance*. (In the press.)

Chan, T. (1999) Pricing contingent claims on stocks driven by Lévy processes. *Annals of Applied Probability* **9**, 504–528.

Clark, P. (1973) A subordinated stochastic process model with finite variance for speculative prices. *Econometrica* **41**, 135–156.

Cont, R. (2001) Empirical properties of asset returns: stylized facts and statistical issues. *Quantitive Finance* **1**, 223–236.

Cont, R., Pooters, M. and Bouchard, J.-P. (1997) Scaling in stock market data: stable laws and beyond. In *Scale Invariance and Beyond (Proceedings of the CNRS Workshop on Scale Invariance, Les Houches, March 1997)* (ed. B. Dubrulle, F. Graner and D. Sornette), pp. 75–85. Springer.

Cox, J., Ingersoll, J. and Ross, S. (1985) A theory of the term structure of interest rates. *Econometrica* **53**, 385–408.

Delbaen, F. and Schachermayer, W. (1994) A general version of the fundamental theorem of asset pricing. *Mathematische Annalen* **300**, 463–520.

Devroye, L. (1986) *Non-Uniform Random Variate Generation*. Springer.

Di Nunno, G. (2001) Stochastic Integral Representations, Stochastic Derivatives and Minimal Variance Hedging. Preprint Series 2001, Pure Mathematics, Department of Mathematics, University of Oslo, no. 19.

Di Nunno, G., Øksendal, B. and Proske F. (2002) White Noise Analysis for Lévy Processes. Preprint Series 2002, Pure Mathematics, Department of Mathematics, University of Oslo, no. 7.

Dritschel, M. and Protter, P. (1999) Complete markets with discontinuous security price. *Finance and Stochastics* **3**, 203–214.

Eberlein, E. and Jacod, J. (1997) On the range of options prices. *Finance and Stochastics* **1**, 131–140.

Eberlein, E. and Keller, U. (1995) Hyperbolic distributions in finance. *Bernoulli* **1**, 281–299.

Eberlein, E. and Prause, K. (1998) The generalized hyperbolic model: financial derivatives and risk measures. FDM Preprint 56, University of Freiburg.

Eberlein, E. and Raible, S. (1999) Term structure models driven by general Lévy processes. *Mathematical Finance* **9**, 31–53.

Eberlein, E. and v. Hammerstein, E. A. (2002) Generalized Hyperbolic and Inverse Gaussian distributions: limiting cases and approximation of processes. FDM Preprint 80, University of Freiburg.

Eberlein, E., Keller, U. and Prause, K. (1998) New insights into smile, mispricing and value at risk: the hyperbolic model. *Journal of Business* **71**, 371–406.

Elliot, R. J. and Kopp, P. E. (1999) *Mathematics of Financial Markets.* Springer.

Emmer, S. and Klüppelberg, C. (2002) Optimal portfolios when stock prices follow an exponential Lévy process. Research Report, Centre of Mathematical Sciences, Munich University of Technology.

Fama, H. (1965) The behavior of stock market prices. *Journal of Business* **38**, 34–105.

Filipović, D. (2001) *Consistency Problems for Heath–Jarrow–Morton Interest Rate Models.* Lecture Notes in Mathematics, vol. 1760. Springer.

Geman, H. (2002) Pure jump Lévy processes for asset price modelling. *Journal of Banking and Finance* **26**, 1297–1316.

Geman, H., Madan, D. and Yor, M. (2001) Time changes for Lévy processes. *Mathematical Finance* **11**, 79–96.

Gerber, H. U. and Shiu, E. S. W. (1994) Option pricing by Esscher-transforms. *Transactions of the Society of Actuaries* **46**, 99–191.

Gerber, H. U. and Shiu, E. S. W. (1996) Actuarial bridges to dynamic hedging and option pricing. *Insurance: Mathematics and Economics* **18**(3), 183–218.

Good, I. J. (1953) The population frequencies of species and the estimation of population parameters. *Biometrika* **40**, 237–260.

Grigelionis, B. (1999) Processes of Meixner type. *Lithuanian Mathematics Journal* **39**(1), 33–41.

Grigelionis, B. (2000) Generalized z-distributions and related stochastic processes. Matematikos Ir Informatikos Institutas Preprintas Nr. 2000-22, Vilnius.

Halgreen, C. (1979) Self-decomposability of the generalized inverse Gaussian and hyperbolic distributions. *Zeitschrift für Wahrscheinlichkeitstheorie und verwandte Gebiete* **47**, 13–18.

Harrison, J. M. and Kreps, D. M. (1979) Martingales and arbitrage in multiperiod securities markets. *Journal of Economic Theory* **20**, 381–408.

Harrison, J. M. and Pliska, S. R. (1981) Martingales and stochastic integrals in the theory of continuous trading. *Stochastic Processes and their Applications* **11**, 215–260.

Haug, E. G. (1998) *The Complete Guide To Option Pricing Formulas.* McGraw-Hill.

Heath, D. R., Jarrow, D. and Morton, A. (1992) Bond pricing and term structure of interest rates: a new methodology for contingent claims valuation. *Econometrica* **60**, 77–105.

Heston, S. L. (1993) A closed form solution for options with stochastic volatility with applications to bonds and currency options. *Review of Financial Studies* **6**, 327–343.

Hougaard, P. (1986) Survival models for hetrogeneous populations derived from stable distributions. *Biometrika* **73**, 387–396.

Hull, J. C. and White, A. (1988) The pricing of options on assets with stochastic volatility. *Journal of Finance* **42**, 281–300.

Hull, J. C. (2000) *Options, Futures and Other Derivatives*, 4th edn. Prentice-Hall.

Hunt, P. J. and Kennedy, J. E. (2000) *Financial Derivatives in Theory and Practice*. John Wiley & Sons, Ltd.

Ingersoll Jr, J. E. (2000) Digital contracts: simple tools for pricing complex derivatives. *Journal of Business* **73**, 67–88.

Itô, K. (1951) Multiple Wiener integral. *Journal of the Mathematical Society of Japan* **3**(1), 157–169.

Jäckel, P. (2002) *Monte Carlo Methods in Finance*. John Wiley & Sons, Ltd.

James, J. and Webber, N. (2000) *Interest Rate Modelling*. John Wiley & Sons, Ltd.

Jarrow, R. A., Jin, X. and Madan, D. B. (1999) The second fundamental theorem of asset pricing. *Mathematical Finance* **9**, 255–273.

Johnk, M. D. (1964) Erzeugung von Betaverteilten und Gammaverteilten Zufallszahlen. *Metrika* **8**, 5–15.

Jørgensen, B. (1982) *Statistical Properties of the Generalized Inverse Gaussian Distribution*. Lecture Notes in Statistics, vol. 9. Springer.

Jurek, Z. J. and Vervaat, W. (1983) An integral representation for selfdecomposable Banach space valued random variables. *Zeitschrift für Wahrscheinlichkeitstheorie und verwandte Gebiete* **62**, 247–262.

Karatzas, I. and Shreve, S. E. (1996) *Brownian Motion and Stochastic Calculus*, 2nd edn. Springer.

Kloeden, P. E. and Platen, E. (1992) *Numerical Solutions of Stochastic Differential Equations*. Springer.

Koekoek, R. and Swarttouw, R. F. (1998) The Askey-scheme of hypergeometric orthogonal polynomials and its q-analogue. Report 98-17, Delft University of Technology.

Koponen, I. (1995) Analytic approach to the problem of convergence of truncated Lévy flights towards the Gaussian stochastic process. *Physical Review* E **52**, 1197–1199.

Kou, S. G. and Wang, H. (2001) Option pricing under a jump diffusion model. Preprint.

Kreps, D. (1981) Arbitrage and equilibrium in economics with infinitely many commodities. *Journal of Mathematical Economics* **8**, 15–35.

Kudryavtsev, O. and Levendorskiĭ, S. (2002) Comparative study of first touch digitals: normal inverse Gaussian vs. Gaussian modelling. Preprint.

Leblanc, B. and Yor, M. (1998) Lévy processes in finance: a remedy to the non-stationarity of continuous martingales. *Finance and Stochastics* **2**, 399–408.

Léon, J. A., Vives, J., Utzet, F. and Solé, J. L. (2002) On Lévy processes, Malliavin calculus and market models with jumps. *Finance and Stochastics* **6**, 197–225.

Lévy, P. (1937) *Théories de L'Addition Aléatories*. Gauthier-Villars, Paris.

Løkka, A. (2001) Martingale representations of functionals of Lévy processes. Preprint Series 2001, Pure Mathematics, Department of Mathematics, University of Oslo, no. 21.

Lukacs, E. (1970) *Characteristic Functions*. Griffin, London.

McKean Jr, H. P. (1965) Appendix: a free boundary problem for the heat equation arising from a problem in mathematical economics. *Industrial Management Review* **6**, 32–39.

Madan, D. B. and Milne, F. (1991) Option pricing with VG martingale components. *Mathematical Finance* **1**(4), 39–55.

Madan, D. B. and Seneta, E. (1987) Chebyshev polynomial approximations and characteristic function estimation. *Journal of the Royal Statistical Society* B **49**(2), 163–169.

Madan, D. B. and Seneta, E. (1990) The VG model for share market returns. *Journal of Business* **63**, 511–524.

Madan, D. B., Carr, P. and Chang, E. C. (1998) The variance Gamma process and option pricing. *European Finance Review* **2**, 79–105.

Marcus, A. H. (1975) Power laws in compartmental analysis. Part I. A unified stochastic model. *Mathematics Biosciences* **23**, 337–350.

Marcus, M. B. (1987) ζ-*Radial Processes and Random Fourier Series*. Memoirs of the American Mathematical Society, vol. 368.

Matacz, A. (1997) Financial modeling and option theory with the truncated Lévy process. University of Sydney Report 97-28.

Merton, R. C. (1973) Theory of rational option pricing. *Bell Journal of Economics and Management Science* **4**, 141–183.

Michael, J. R., Schucany, W. R. and Haas, R. W. (1976) Generating random variates using transformations with multiple roots. *The American Statistician* **30**, 88–90.

Mordecki, E. (2002) Optimal stopping and perpetual options for Lévy processes. *Finance and Stochastics* **6**, 473–493.

Mordecki, E. and Moreira, W. (2002) Russian options for a diffusion with negative jumps. *Publicaciones Matemáticas del Uruguay* **9**. (In the press.)

Neuts, M. (1981) *Matrix-Geometric Solutions in Stochastic Models*. Johns Hopkins University Press.

Nicolato, E. and Venardos, E. (2003) Option pricing in stochastic volatility models of Ornstein–Uhlenbeck type. *Mathematical Finance*. (In the press.)

Nualart, D. and Schoutens W. (2000) Chaotic and predictable representations for Lévy processes. *Stochastic Processes and their Applications* **90**(1), 109–122.

Nualart, D. and Schoutens W. (2001) Backwards stochastic differential equations and Feynman–Kac formula for Lévy processes, with applications in finance. *Bernoulli* **7**, 761–776.

Øksendal, B. and Proske, F. (2002) White noise of Poisson random measures. Preprint Series 2002, Pure Mathematics, Department of Mathematics, University of Oslo, no. 12.

Pecherskii, E. A. and Rogozin, B. A. (1969) On joint distributions of random variables associated with fluctuations of a process with independent increments. *Theory of Probability and Its Applications* **14**, 410–423.

Pitman, J. and Yor, M. (2000) Infinitely divisible laws associated with hyperbolic functions. Prépublications du Laboratoire de Probabilités et Modèles Aléatoires, vol. 616. Universités de Paris 6 and Paris 7, Paris.

Pollard, D. (1984) *Convergence of Stochastic Processes*. Springer Series in Statistics. Springer.

Prause, K. (1999) The Generalized Hyperbolic model: estimation, financial derivatives, and risk measures. PhD thesis, Freiburg.

Protter, Ph. (1990) *Stochastic Integration and Differential Equations*. Springer.

Protter, Ph. (2001) A partial introduction to financial asset pricing theory. *Stochastic Processes and Their Applications* **91**, 169–203.

Raible, S. (2000) Lévy processes in finance: theory, numerics, and empirical facts. PhD thesis, Freiburg.

Rebonato, R. (1996) *Interest-Rate Option Models.* John Wiley & Sons, Ltd.

Rosiński, J. (1991) On a class of infinitely divisible processes represented as mixtures of Gaussian processes. In *Stable Processes and Related Topics* (ed. S. Cambanis, G. Samorodnitsky and M. S. Taqqu), pp. 27–41. Birkhäuser, Basel.

Rosiński, J. (2001) Series representations of Lévy processes from the perspective of point processes. In *Lévy Processes – Theory and Applications* (ed. O. E. Barndorff-Nielsen, T. Mikosch and S. Resnick), pp. 401–415. Birkhäuser, Boston.

Rosiński, J. (2002) Tempered stable processes. In *Miniproceedings of 2nd MaPhySto Conference on Lévy Processes: Theory and Applications* (ed. O. E. Barndorff-Nielsen), pp. 215–220.

Rydberg, T. (1996a) The normal inverse Gaussian Lévy process: simulations and approximation. Research Report 344, Department of Theoretical Statistics, Aarhus University.

Rydberg, T. (1996b) Generalized hyperbolic diffusions with applications towards finance. Research Report 342, Department of Theoretical Statistics, Aarhus University.

Rydberg, T. (1997a) A note on the existence of unique equivalent martingale measures in a Markovian setting. *Finance and Stochastics* **1**, 251–257.

Rydberg, T. (1997b) The normal inverse Gaussian Lévy process: simulations and approximation. *Communications in Statistics: Stochastic Models* **13**, 887–910.

Rydberg, T. (1998) Some modelling results in the area of interplay between statistics, mathematical finance, insurance and econometrics. PhD thesis, University of Aarhus.

Samuelson, P. (1965) Rational theory of warrant pricing. *Industrial Management Review* **6**, 13–32.

Sato, K. (1999) *Lévy Processes and Infinitely Divisible Distributions.* Cambridge Studies in Advanced Mathematics, vol. 68. Cambridge University Press.

Sato, K. and Yamazato, M. (1982) Stationary processes of Ornstein–Uhlenbeck type. In *Probability Theory and Mathematical Statistics* (ed. K. Itô and J. V. Prohorov). Lecture Notes in Mathematics, vol. 1021. Springer.

Sato, K., Watanabe, T. and Yamazato, M. (1994) Recurrence conditions for multidimensional processes of Ornstein–Uhlenbeck type. *Journal of the Mathematical Society of Japan* **46**, 245–265.

Schoutens, W. (2000) *Stochastic Processes and Orthogonal Polynomials.* Lecture Notes in Statistics, vol. 146. Springer.

Schoutens, W. (2001) The Meixner process in finance. EURANDOM Report 2001-002. EURANDOM, Eindhoven.

Schoutens, W. (2002) Meixner processes: theory and applications in finance. EURANDOM Report 2002-004. EURANDOM, Eindhoven.

Schoutens, W. and Teugels, J. L. (1998) Lévy processes, polynomials and martingales. *Communications in Statistics: Stochastic Models* **14**, 335–349.

Shaw, W. T. (1998) *Modelling Financial Derivatives with Mathematica.* Cambridge University Press.

Shepp, L. and Shiryaev, A. N. (1993) The Russian option: reduced regret. *Annals of Applied Probability* **3**, 603–631.

Shepp, L. and Shiryaev, A. N. (1994) A new look at the pricing of the Russian option. *Theory of Probability and Its Applications* **39**, 103–120.

Shiryaev, A. N. (1999) *Essentials of Stochastic Finance.* World Scientific.

Sichel, H. S. (1974) On a distribution representing sentence-length in written prose. *Journal of the Royal Statistical Society* A **137**, 25–34.

Sichel, H. S. (1975) On a distribution law for word frequencies. *Journal of the American Statistical Society Association* **70**, 542–547.

Silverman, B. W. (1986) *Density Estimation for Statistics and Data Analysis*. Chapman and Hall, London.

Tompkins, R. and Hubalek, F. (2000) On closed form solutions for pricing options with jumping volatility. Unpublished paper, Technical University, Vienna.

Tweedie, M. C. K. (1947) Functions of a statistical variate with given means, with special reference to Laplacian distributions. *Proceedings of the Cambridge Philosophical Society* **43**, 41–49.

Tweedie, M. C. K. (1984) An index which distinguishes between some important exponential families. In *Statistics: Applications and New Directions: Proc. Indian Statistical Institute Golden Jubilee International Conference* (ed. J. Ghosh and J. Roy), pp. 579–604.

Wald (1947) *Sequential Analysis*. John Wiley & Sons, Ltd, New York.

Wise, M. G. (1975) Skew distributions in biomedicine including some with negative powers of time. In *Statistical Distributions in Scientific Work*, Vol. 2: *Model Building and Model Selection* (ed. G. P. Patil *et al.*), pp. 241–262. Dordrecht, Reidel.

Wolfe, S. J. (1982) On a continuous analogue of the stochastic difference equation $\rho X_{n+1} + B_n$. *Stochastic Processes and Their Applications* **12**, 301–312.

Yor, M. (1992) *Some Aspects of Brownian Motion*, Part I: *Some Special Functionals*. Lectures in Mathematics ETH Zürich. Birkhäuser, Berlin.

Yor, M. and Nguyen, L. (2001) Wiener–Hopf Factorization and the Pricing of Barrier and Lookback Options under General Lévy Processes. Prépublications du Laboratoire de Probabilités et Modèles Aléatoires, vol. 640. Universités de Paris 6 et Paris 7, Paris.

Index

AAE, 7
APE, 7
arbitrage, 9
ARPE, 7

Background Driving Lévy Process,
   48
bank account, 7, 137
Barndorff-Nielsen–Shephard model,
   see also BNS model
barrier option, 4, 119
   Black–Scholes prices, 121
   down-and-in, 120
   down-and-out, 120
   Lévy market prices, 123
   Lévy SV prices, 131
   up-and-in, 120
   up-and-out, 42, 120
BDLP, *see also* Background Driving
   Lévy Process
Berman's Gamma generator, 109
Bessel functions, 147
   modified, 148
binary barrier option, 123
binary option, 123
Black–Scholes model, 28
   completeness, 30
   imperfections, 33
   option pricing, 30
   partial differential equation, 31
   risk-neutral setting, 30

stochastic volatility, 85
BNS model, 85
   simulation, 117
bond
   zero-coupon, 135
bourse, 1
Brownian motion, 24, 153
   definition, 25
   history, 24
   properties, 26
   simulation, 101

càdlàg, 14
calibration
   Black–Scholes model, 39
   BNS models, 97
   Lévy models, 82
   Lévy SV models, 98
   stochastic volatility models, 97
CBOE, 5
CGMY distribution, 60, 152
CGMY process, 60, 153
characteristic exponent, 44
characteristic function, 15, 151
   option pricing, 20
χ^2-test, 37, 75
CIR process, 89, 137
   simulation, 116
compound Poisson process, 51, 103
contingent claim, 3
   history, 5
controls variates, 129
Cox–Ingersoll–Ross process, *see
   also* CIR process
cumulant characteristic function, 16

Lévy Processes in Finance   W. Schoutens
© 2003 John Wiley & Sons, Ltd
ISBN: 0-470-85156-2

cumulant function, 16

D-OU process, 48
delta, 20, 31
density estimation, 35
density function
   option pricing, 19
derivative, 3
   types, 3
diffusion component, 76
discounting factor, 19, 135
distribution
   CGMY, 60, 152
   Gamma, 52, 108, 151
   Generalized z, 64, 152
   Generalized Hyperbolic, 65, 152
   Generalized Inverse Gaussian, 54,
      151, 152
   Hyperbolic, 66, 152
   infinitely divisible, 44
   Inverse Gaussian, 53, 111, 151
   Meixner, 62, 152
   Normal, 23, 152
   Normal Inverse Gaussian, 59, 152
   Poisson, 50, 151
   Tempered Stable, 56, 151
   Variance Gamma, 57, 152
dividends, 2, 21
drift coefficient, 45
drift term, 67

equivalent martingale measure, 17
   BNS model, 86
   Esscher transform, 77
   existence, 17
   Gaussian HJM model, 142
   Lévy HJM model, 142
   mean correcting, 79, 91
   uniqueness, 18
Esscher transform, 77, 124
EURIBOR, 136
excess kurtosis, 34

fat tails, 34

filtration, 12
financial asset, 1
finite activity, 45
finite variation, 14, 45
forwards, 3
futures, 3

gamma distribution, 52, 151
gamma process, 52, 153
   simulation, 108
gamma–OU process, 68, 87, 90
   simulation, 114
Gaussian HJM model, 138
Gaussian kernel density estimator, 35
Generalized z distribution, 64, 152
Generalized Hyperbolic distribution,
   65, 152
Generalized Hyperbolic process, 65,
   153
generalized hypergeometric series,
   149
Generalized Inverse Gaussian
   distribution, 54, 151, 152
Generalized Inverse Gaussian
   process, 54, 153
geometric Brownian motion, 27
GH distribution, *see also* Generalized
   Hyperbolic distribution
GH process, *see also* Generalized
   Hyperbolic process
GIG distribution, *see also*
   Generalized Inverse Gaussian
   distribution
GIG process, *see also* Generalized
   Inverse Gaussian process
GZ distribution, *see also* Generalized
   z distribution

hedging, 18
Hermite polynomial, 64, 149
historical volatility, 38
HJM drift conditions, 139
hyperbolic distribution, 152
hyperbolic process, 66

IG distribution, *see also* Inverse
    Gaussian distribution
IG process, *see also* Inverse Gaussian
    process
IG–OU process, 69, 88, 91
    simulation, 115
implied volatility, 39
implied volatility models, 41
index, 1
    S&P 500, 1
infinite variation, 14
    Brownian motion, 26
    Lévy process, 45
infinitely divisible, 44
instantaneous forward rate, 138
    HJM model, 139
integrated Ornstein–Uhlenbeck
    process, *see also* intOU process
integrated OU process, *see also*
    intOU process
interest rate, 7, 135
    short rate, 136
    term structure, 136
intOU process, 49
Inverse Gaussian distribution, 53, 151
Inverse Gaussian process, 53, 153
    simulation, 111
Itô calculus, 16

Johnk's Gamma generator, 108

kernel density estimator, 35
kurtosis, 34

Lévy HJM model, 141
Lévy market model, 76
Lévy measure, 154
Lévy process, 44, 151
    Brownian motion, 24, 153
    CGMY process, 60, 153
    compound Poisson
        approximation, 103
    compound Poisson process, 51
    drift coefficient, 45

finite activity, 45
finite variation, 45
Gamma process, 52, 153
Generalized Hyperbolic process,
    65, 153
Generalized Inverse Gaussian
    process, 54, 153
Hyperbolic process, 66
infinite variation, 45
Inverse Gaussian process, 53, 153
Meixner process, 62, 153
Normal Inverse Gaussian process,
    59, 153
Poisson process, 50, 153
predictable representation
    property, 46
simulation, 102
Tempered Stable process, 56, 153
Variance Gamma process, 57, 153
Lévy SV market model, 91
Lévy triplet, 45, 103, 153
Lévy, Paul, 46
Lévy measure, 45
Lévy–Khintchine formula, 44
leptokurtic, 35
leverage effect, 86
log density, 36, 141
log return, 3
lookback option, 4, 119
    Black–Scholes prices, 121, 122
    Lévy market prices, 123

MacDonald function, 148
market incompleteness, 77
Markov process, 13
martingale, 14
    Brownian motion, 26
    equivalent martingale measure, 17
    Hermite polynomial, 64
    Meixner–Pollaczek polynomial,
        64
    predictable representation
        property, 18
maximum-likelihood estimator, 74

mean-correcting martingale measure,
        79
measures of fit, 7
Meixner distribution, 62, 152
Meixner process, 62, 153
Meixner–OU process, 71
Meixner–Pollaczek polynomial, 64,
        150
mesokurtic, 35
MLE, *see also* maximum likelihood
        estimator
modelling assumptions, 7
        bank account, 7
        no-arbitrage, 9
        risky asset, 8
modified Bessel functions, 148
moment-generating function, 16
money-market account, 137
Monte Carlo pricing, 127
        BNS models, 127
        Lévy SV models, 128
Monte Carlo simulation, 127
multi-factor Gaussian HJM model,
        144
multi-factor Lévy HJM model, 145
multi-factor model, 144

NIG distribution, *see also* Normal
        Inverse Gaussian distribution
NIG process, *see also* Normal
        Inverse Gaussian process
NIG–OU process, 71
no free lunch with vanishing risk, 17
no-arbitrage, 9, 17
        no free lunch with vanishing risk,
        17
Normal distribution, 23, 152
normal Inverse Gaussian distribution,
        59, 152
normal Inverse Gaussian process, 59,
        153
        simulation, 113

option pricing, 19

Black–Scholes, 31
Gaussian HJM model, 143
inversion of distribution function
        transform, 20
inversion of the modified call
        price, 20
Lévy HJM model, 143
Monte Carlo methods, 127
through the characteristic
        function, 20
through the density function, 19
options, 3
advantages and disadvantages, 4
American, 4
Asian, 4
at the money, 4
barrier, 4, 119
binary, 123
binary barrier, 123
bond, 142
call, 4
delta, 20, 31
European, 4
exotic, 4, 119
history, 5
in the money, 4
lookback, 4, 119
out the money, 4
perpetual American, 125
perpetual Russian, 126
plain vanilla, 4
pricing, 19
put, 4
touch-and-out, 126
types, 4
vega, 41
Ornstein–Uhlenbeck process, *see
        also* OU process
OU process, 47, 90
        Gamma–OU process, 68
        IG–OU process, 69
        integrated, 49
        Meixner–OU process, 71
        NIG–OU process, 71

simulation, 107
tail mass function, 49
TS–OU process, 70
OU-$\tilde{D}$ process, 48
over-the-counter (OTC), 5

P-value, 37, 75
partial differential integral equation,
    82
payoff function, 4
PDIE, *see also* partial differential
    integral equation
perpetual American option, 125
perpetual Russian option, 126
platykurtic, 35
Poisson distribution, 50, 151
Poisson process, 50, 153
    predictable representation
        property, 51
    simulation, 102
predictable representation property,
    18
    Brownian motion, 30
    Lévy process, 46
    Poisson process, 51
probability space, 11
    filtered, 12
PRP, *see also* predictable
    representation property
put–call parity, 9
    dividends, 21

random number generator
    Berman's Gamma generator, 109
    Gamma distribution, 108
    IG distribution, 111
    Johnk's Gamma generator, 108
return, 3
    log return, 3
RMSE, 7

S&P 500 Index, 2
    option prices dataset, 6, 155
SDE, *see also* stochastic differential
    equation

self-decomposability, 47
semi-heavy tails, 36
short rate, 136
    CIR model, 137
    under HJM model, 139
simulation, 101
    BNS model, 117
    Brownian motion, 101
    CIR process, 116
    Gamma process, 108
    Gamma–OU process, 114
    IG–OU process, 115
    Inverse Gaussian process, 111
    Lévy process, 102
    Normal Inverse Gaussian process,
        113
    OU process, 107
    Poisson process, 102
    Tempered Stable process, 111
    Variance Gamma process, 109
skewness, 34
standard error, 127
stochastic differential equation, 16
    Black–Scholes, 28
    CIR process, 89, 137
    geometric Brownian motion, 28
    OU process, 48
stochastic integral, 16
stochastic process, 12
    adapted, 12
    Brownian motion, 24
    finite variation, 14
    geometric Brownian motion, 27
    infinite variation, 14
    Lévy process, 44
    Markov process, 13
    martingale, 14
    OU process, 47
    predictable, 12
stochastic volatility, 38
stock, 1
Stock Exchange, 1
strike price, 4
stylized features of financial data, 33

swaps, 3

tail mass function, 49
    Gamma–OU process, 69
    inverse, 50, 114, 115
Tempered Stable distribution, 56, 151
Tempered Stable process, 56, 153
    simulation, 111, 115
term structure of interest rates, 136
time change, 88, 90, 91
touch-and-out option, 126
triplet of Lévy characteristics, 45,
    103, 153
TS distribution, *see also* Tempered
    Stable distribution
TS process, *see also* Tempered Stable
    process
TS–OU process, 70

variance Gamma distribution, 57, 152
variance Gamma process, 57, 153
    simulation, 109
variance reduction, 129
vega, 41

VG distribution, *see also* Variance
    Gamma distribution
VG process, *see also* Variance
    Gamma process
volatility
    clusters, 38
    historical, 38
    Ho–Lee structure, 140
    implied, 39
    stochastic, 38
    surface, 41
    Vasicek structure, 140
volatility clusters, 38
volatility structure, 140
    Ho–Lee, 140
    Vasicek, 140
volatility surface, 41

yield curve, 136

zero-coupon bond, 135
    Gaussian HJM model, 140
    Lévy HJM model, 142

WILEY SERIES IN PROBABILITY AND STATISTICS

ESTABLISHED BY WALTER A. SHEWHART AND SAMUEL S. WILKS

Editors: *David J. Balding, Peter Bloomfield, Noel A. C. Cressie,*
Nicholas I. Fisher, Iain M. Johnstone, J. B. Kadane, Louise M. Ryan,
David W. Scott, Adrian F. M. Smith, Jozef L. Teugels
Editors Emeriti: *Vic Barnett, J. Stuart Hunter, David G. Kendall*

The *Wiley Series in Probability and Statistics* is well established and authoritative. It covers many topics of current research interest in both pure and applied statistics and probability theory. Written by leading statisticians and institutions, the titles span both state-of-the-art developments in the field and classical methods.

Reflecting the wide range of current research in statistics, the series encompasses applied, methodological and theoretical statistics, ranging from applications and new techniques made possible by advances in computerized practice to rigorous treatment of theoretical approaches.

This series provides essential and invaluable reading for all statisticians, whether in academia, industry, government, or research.

ABRAHAM and LEDOLTER · Statistical Methods for Forecasting
AGRESTI · Analysis of Ordinal Categorical Data
AGRESTI · An Introduction to Categorical Data Analysis
AGRESTI · Categorical Data Analysis, *Second Edition*
ANDĚL · Mathematics of Chance
ANDERSON · An Introduction to Multivariate Statistical Analysis, *Second Edition*
*ANDERSON · The Statistical Analysis of Time Series
ANDERSON, AUQUIER, HAUCK, OAKES, VANDAELE and WEISBERG ·
Statistical Methods for Comparative Studies
ANDERSON and LOYNES · The Teaching of Practical Statistics
ARMITAGE and DAVID (editors) · Advances in Biometry
ARNOLD, BALAKRISHNAN and NAGARAJA · Records
*ARTHANARI and DODGE · Mathematical Programming in Statistics
*BAILEY · The Elements of Stochastic Processes with Applications to the Natural
Sciences
BALAKRISHNAN and KOUTRAS · Runs and Scans with Applications
BARNETT · Comparative Statistical Inference, *Third Edition*
BARNETT and LEWIS · Outliers in Statistical Data, *Third Edition*

---

*Now available in a lower-priced paperback edition in the Wiley Classics Library.

BARTOSZYNSKI and NIEWIADOMSKA-BUGAJ · Probability and Statistical Inference

BASILEVSKY · Statistical Factor Analysis and Related Methods: Theory and Applications

BASU and RIGDON · Statistical Methods for the Reliability of Repairable Systems

BATES and WATTS · Nonlinear Regression Analysis and Its Applications

BECHHOFER, SANTNER and GOLDSMAN · Design and Analysis of Experiments for Statistical Selection, Screening, and Multiple Comparisons

BELSLEY · Conditioning Diagnostics: Collinearity and Weak Data in Regression

BELSLEY, KUH and WELSCH · Regression Diagnostics: Identifying Influential Data and Sources of Collinearity

BENDAT and PIERSOL · Random Data: Analysis and Measurement Procedures, *Third Edition*

BERRY, CHALONER and GEWEKE · Bayesian Analysis in Statistics and Econometrics: Essays in Honour of Arnold Zellner

BERNARDO and SMITH · Bayesian Theory

BHAT and MILLER · Elements of Applied Stochastic Processes, *Third Edition*

BHATTACHARYA and JOHNSON · Statistical Concepts and Methods

BHATTACHARYA and WAYMIRE · Stochastic Processes with Applications

BILLINGSLEY · Convergence of Probability Measures, *Second Edition*

BILLINGSLEY · Probability and Measure, *Third Edition*

BIRKES and DODGE · Alternative Methods of Regression

BLISCHKE AND MURTHY · Reliability: Modeling, Prediction, and Optimization

BLOOMFIELD · Fourier Analysis of Time Series: An Introduction, *Second Edition*

BOLLEN · Structural Equations with Latent Variables

BOROVKOV · Ergodicity and Stability of Stochastic Processes

BOULEAU · Numerical Methods for Stochastic Processes

BOX · Bayesian Inference in Statistical Analysis

BOX · R. A. Fisher, the Life of a Scientist

BOX and DRAPER · Empirical Model-Building and Response Surfaces

*BOX and DRAPER · Evolutionary Operation: A Statistical Method for Process Improvement

BOX, HUNTER and HUNTER · Statistics for Experimenters: An Introduction to Design, Data Analysis, and Model Building

BOX and LUCEÑO · Statistical Control by Monitoring and Feedback Adjustment

BRANDIMARTE · Numerical Methods in Finance: A MATLAB-Based Introduction

BROWN and HOLLANDER · Statistics: A Biomedical Introduction

BRUNNER, DOMHOF and LANGER · Nonparametric Analysis of Longitudinal Data in Factorial Experiments

BUCKLEW · Large Deviation Techniques in Decision, Simulation, and Estimation

CAIROLI and DALANG · Sequential Stochastic Optimization

---

*Now available in a lower priced paperback edition in the Wiley Classics Library.

CHAN · Time Series: Applications to Finance

CHATTERJEE and HADI · Sensitivity Analysis in Linear Regression

CHATTERJEE and PRICE · Regression Analysis by Example, *Third Edition*

CHERNICK · Bootstrap Methods: A Practitioner's Guide

CHILÈS and DELFINER · Geostatistics: Modeling Spatial Uncertainty

CHOW and LIU · Design and Analysis of Clinical Trials: Concepts and Methodologies

CLARKE and DISNEY · Probability and Random Processes: A First Course with Applications, *Second Edition*

*COCHRAN and COX · Experimental Designs, *Second Edition*

CONGDON · Applied Bayesian Modelling

CONGDON · Bayesian Statistical Modelling

CONOVER · Practical Nonparametric Statistics, *Second Edition*

COOK · Regression Graphics

COOK and WEISBERG · Applied Regression Including Computing and Graphics

COOK and WEISBERG · An Introduction to Regression Graphics

CORNELL · Experiments with Mixtures, Designs, Models, and the Analysis of Mixture Data, *Third Edition*

COVER and THOMAS · Elements of Information Theory

COX · A Handbook of Introductory Statistical Methods

*COX · Planning of Experiments

CRESSIE · Statistics for Spatial Data, *Revised Edition*

CSÖRGÖ and HORVÁTH · Limit Theorems in Change Point Analysis

DANIEL · Applications of Statistics to Industrial Experimentation

DANIEL · Biostatistics: A Foundation for Analysis in the Health Sciences, *Sixth Edition*

*DANIEL · Fitting Equations to Data: Computer Analysis of Multifactor Data, *Second Edition*

DAVID · Order Statistics, *Second Edition*

*DEGROOT, FIENBERG and KADANE · Statistics and the Law

DEL CASTILLO · Statistical Process Adjustment for Quality Control

DENISON, HOLMES, MALLICK and SMITH · Bayesian Methods for Nonlinear Classification and Regression

DETTE and STUDDEN · The Theory of Canonical Moments with Applications in Statistics, Probability, and Analysis

DEY and MUKERJEE · Fractional Factorial Plans

DILLON and GOLDSTEIN · Multivariate Analysis: Methods and Applications

DODGE · Alternative Methods of Regression

*DODGE and ROMIG · Sampling Inspection Tables, *Second Edition*

*DOOB · Stochastic Processes

DOWDY and WEARDEN · Statistics for Research, *Second Edition*

DRAPER and SMITH · Applied Regression Analysis, *Third Edition*

---

*Now available in a lower priced paperback edition in the Wiley Classics Library.

DRYDEN and MARDIA · Statistical Shape Analysis

DUDEWICZ and MISHRA · Modern Mathematical Statistics

DUNN and CLARK · Applied Statistics: Analysis of Variance and Regression, *Second Edition*

DUNN and CLARK · Basic Statistics: A Primer for the Biomedical Sciences, *Third Edition*

DUPUIS and ELLIS · A Weak Convergence Approach to the Theory of Large Deviations

*ELANDT-JOHNSON and JOHNSON · Survival Models and Data Analysis

ETHIER and KURTZ · Markov Processes: Characterization and Convergence

EVANS, HASTINGS and PEACOCK · Statistical Distributions, *Third Edition*

FELLER · An Introduction to Probability Theory and Its Applications, Volume I, *Third Edition*, Revised; Volume II, *Second Edition*

FISHER and VAN BELLE · Biostatistics: A Methodology for the Health Sciences

*FLEISS · The Design and Analysis of Clinical Experiments

FLEISS · Statistical Methods for Rates and Proportions, *Second Edition*

FLEMING and HARRINGTON · Counting Processes and Survival Analysis

FULLER · Introduction to Statistical Time Series, *Second Edition*

FULLER · Measurement Error Models

GALLANT · Nonlinear Statistical Models

GHOSH, MUKHOPADHYAY and SEN · Sequential Estimation

GIFI · Nonlinear Multivariate Analysis

GLASSERMAN and YAO · Monotone Structure in Discrete-Event Systems

GNANADESIKAN · Methods for Statistical Data Analysis of Multivariate Observations, *Second Edition*

GOLDSTEIN and LEWIS · Assessment: Problems, Development, and Statistical Issues

GREENWOOD and NIKULIN · A Guide to Chi-Squared Testing

GROSS and HARRIS · Fundamentals of Queueing Theory, *Third Edition*

*HAHN · Statistical Models in Engineering

HAHN and MEEKER · Statistical Intervals: A Guide for Practitioners

HALD · A History of Probability and Statistics and Their Applications Before 1750

HALD · A History of Mathematical Statistics from 1750 to 1930

HAMPEL · Robust Statistics: The Approach Based on Influence Functions

HANNAN and DEISTLER · The Statistical Theory of Linear Systems

HEIBERGER · Computation for the Analysis of Designed Experiments

HEDAYAT and SINHA · Design and Inference in Finite Population Sampling

HELLER · MACSYMA for Statisticians

HINKELMAN and KEMPTHORNE: · Design and Analysis of Experiments, Volume 1: Introduction to Experimental Design

HOAGLIN, MOSTELLER and TUKEY · Exploratory Approach to Analysis of Variance

*Now available in a lower priced paperback edition in the Wiley Classics Library.

HOAGLIN, MOSTELLER and TUKEY · Exploring Data Tables, Trends and Shapes
*HOAGLIN, MOSTELLER and TUKEY · Understanding Robust and Exploratory Data Analysis
HOCHBERG and TAMHANE · Multiple Comparison Procedures
HOCKING · Methods and Applications of Linear Models: Regression and the Analysis of Variables
HOEL · Introduction to Mathematical Statistics, *Fifth Edition*
HOGG and KLUGMAN · Loss Distributions
HOLLANDER and WOLFE · Nonparametric Statistical Methods, *Second Edition*
HOSMER and LEMESHOW · Applied Logistic Regression, *Second Edition*
HOSMER and LEMESHOW · Applied Survival Analysis: Regression Modeling of Time to Event Data
HØYLAND and RAUSAND · System Reliability Theory: Models and Statistical Methods
HUBER · Robust Statistics
HUBERTY · Applied Discriminant Analysis
HUNT and KENNEDY · Financial Derivatives in Theory and Practice
HUSKOVA, BERAN and DUPAC · Collected Works of Jaroslav Hajek—with Commentary
IMAN and CONOVER · A Modern Approach to Statistics
JACKSON · A User's Guide to Principle Components
JOHN · Statistical Methods in Engineering and Quality Assurance
JOHNSON · Multivariate Statistical Simulation
JOHNSON and BALAKRISHNAN · Advances in the Theory and Practice of Statistics: A Volume in Honor of Samuel Kotz
JUDGE, GRIFFITHS, HILL, LÜTKEPOHL and LEE · The Theory and Practice of Econometrics, *Second Edition*
JOHNSON and KOTZ · Distributions in Statistics
JOHNSON and KOTZ (editors) · Leading Personalities in Statistical Sciences: From the Seventeenth Century to the Present
JOHNSON, KOTZ and BALAKRISHNAN · Continuous Univariate Distributions, Volume 1, *Second Edition*
JOHNSON, KOTZ and BALAKRISHNAN · Continuous Univariate Distributions, Volume 2, *Second Edition*
JOHNSON, KOTZ and BALAKRISHNAN · Discrete Multivariate Distributions
JOHNSON, KOTZ and KEMP · Univariate Discrete Distributions, *Second Edition*
JUREČKOVÁ and SEN · Robust Statistical Procedures: Asymptotics and Interrelations
JUREK and MASON · Operator-Limit Distributions in Probability Theory
KADANE · Bayesian Methods and Ethics in a Clinical Trial Design
KADANE AND SCHUM · A Probabilistic Analysis of the Sacco and Vanzetti Evidence

---

*Now available in a lower priced paperback edition in the Wiley Classics Library.

KALBFLEISCH and PRENTICE · The Statistical Analysis of Failure Time Data, *Second Edition*

KASS and VOS · Geometrical Foundations of Asymptotic Inference

KAUFMAN and ROUSSEEUW · Finding Groups in Data: An Introduction to Cluster Analysis

KEDEM and FOKIANOS · Regression Models for Time Series Analysis

KENDALL, BARDEN, CARNE and LE · Shape and Shape Theory

KHURI · Advanced Calculus with Applications in Statistics, *Second Edition*

KHURI, MATHEW and SINHA · Statistical Tests for Mixed Linear Models

KLUGMAN, PANJER and WILLMOT · Loss Models: From Data to Decisions

KLUGMAN, PANJER and WILLMOT · Solutions Manual to Accompany Loss Models: From Data to Decisions

KOTZ, BALAKRISHNAN and JOHNSON · Continuous Multivariate Distributions, Volume 1, *Second Edition*

KOTZ and JOHNSON (editors) · Encyclopedia of Statistical Sciences: Volumes 1 to 9 with Index

KOTZ and JOHNSON (editors) · Encyclopedia of Statistical Sciences: Supplement Volume

KOTZ, READ and BANKS (editors) · Encyclopedia of Statistical Sciences: Update Volume 1

KOTZ, READ and BANKS (editors) · Encyclopedia of Statistical Sciences: Update Volume 2

KOVALENKO, KUZNETZOV and PEGG · Mathematical Theory of Reliability of Time-Dependent Systems with Practical Applications

LACHIN · Biostatistical Methods: The Assessment of Relative Risks

LAD · Operational Subjective Statistical Methods: A Mathematical, Philosophical, and Historical Introduction

LAMPERTI · Probability: A Survey of the Mathematical Theory, *Second Edition*

LANGE, RYAN, BILLARD, BRILLINGER, CONQUEST and GREENHOUSE · Case Studies in Biometry

LARSON · Introduction to Probability Theory and Statistical Inference, *Third Edition*

LAWLESS · Statistical Models and Methods for Lifetime Data

LAWSON · Statistical Methods in Spatial Epidemiology

LE · Applied Categorical Data Analysis

LE · Applied Survival Analysis

LEE and WANG · Statistical Methods for Survival Data Analysis, *Third Edition*

LEPAGE and BILLARD · Exploring the Limits of Bootstrap

LEYLAND and GOLDSTEIN (editors) · Multilevel Modelling of Health Statistics

LIAO · Statistical Group Comparison

LINDVALL · Lectures on the Coupling Method

LINHART and ZUCCHINI · Model Selection

LITTLE and RUBIN · Statistical Analysis with Missing Data, *Second Edition*

LLOYD · The Statistical Analysis of Categorical Data

McCULLOCH and SEARLE · Generalized, Linear, and Mixed Models

McFADDEN · Management of Data in Clinical Trials

McLACHLAN · Discriminant Analysis and Statistical Pattern Recognition

McLACHLAN and KRISHNAN · The EM Algorithm and Extensions

McLACHLAN and PEEL · Finite Mixture Models

McNEIL · Epidemiological Research Methods

MAGNUS and NEUDECKER · Matrix Differential Calculus with Applications in Statistics and Econometrics, *Revised Edition*

MALLER and ZHOU · Survival Analysis with Long Term Survivors

MALLOWS · Design, Data, and Analysis by Some Friends of Cuthbert Daniel

MANN, SCHAFER and SINGPURWALLA · Methods for Statistical Analysis of Reliability and Life Data

MANTON, WOODBURY and TOLLEY · Statistical Applications Using Fuzzy Sets

MARDIA and JUPP · Directional Statistics

MASON, GUNST and HESS · Statistical Design and Analysis of Experiments with Applications to Engineering and Science

MEEKER and ESCOBAR · Statistical Methods for Reliability Data

MEERSCHAERT and SCHEFFLER · Limit Distributions for Sums of Independent Random Vectors: Heavy Tails in Theory and Practice

*MILLER · Survival Analysis, *Second Edition*

MONTGOMERY, PECK and VINING · Introduction to Linear Regression Analysis, *Third Edition*

MORGENTHALER and TUKEY · Configural Polysampling: A Route to Practical Robustness

MUIRHEAD · Aspects of Multivariate Statistical Theory

MÜLLER and STOYAN · Comparison Methods for Stochastic Models and Risks

MURRAY · X-STAT 2.0 Statistical Experimentation, Design Data Analysis, and Nonlinear Optimization

MYERS and MONTGOMERY · Response Surface Methodology: Process and Product Optimization Using Designed Experiments, *Second Edition*

MYERS, MONTGOMERY and VINING · Generalized Linear Models. With Applications in Engineering and the Sciences

NELSON · Accelerated Testing, Statistical Models, Test Plans, and Data Analyses

NELSON · Applied Life Data Analysis

NEWMAN · Biostatistical Methods in Epidemiology

OCHI · Applied Probability and Stochastic Processes in Engineering and Physical Sciences

OKABE, BOOTS, SUGIHARA and CHIU · Spatial Tesselations: Concepts and Applications of Voronoi Diagrams, *Second Edition*

OLIVER and SMITH · Influence Diagrams, Belief Nets and Decision Analysis

PANKRATZ · Forecasting with Dynamic Regression Models

---

*Now available in a lower priced paperback edition in the Wiley Classics Library.

PANKRATZ · Forecasting with Univariate Box-Jenkins Models: Concepts and Cases

*PARZEN · Modern Probability Theory and Its Applications

PEÑA, TIAO and TSAY · A Course in Time Series Analysis

PIANTADOSI · Clinical Trials: A Methodologic Perspective

PORT · Theoretical Probability for Applications

POURAHMADI · Foundations of Time Series Analysis and Prediction Theory

PRESS · Bayesian Statistics: Principles, Models, and Applications

PRESS and TANUR · The Subjectivity of Scientists and the Bayesian Approach

PUKELSHEIM · Optimal Experimental Design

PURI, VILAPLANA and WERTZ · New Perspectives in Theoretical and Applied Statistics

PUTERMAN · Markov Decision Processes: Discrete Stochastic Dynamic Programming

*RAO · Linear Statistical Inference and Its Applications, *Second Edition*

RENCHER · Linear Models in Statistics

RENCHER · Methods of Multivariate Analysis, *Second Edition*

RENCHER · Multivariate Statistical Inference with Applications

RIPLEY · Spatial Statistics

RIPLEY · Stochastic Simulation

ROBINSON · Practical Strategies for Experimenting

ROHATGI and SALEH · An Introduction to Probability and Statistics, *Second Edition*

ROLSKI, SCHMIDLI, SCHMIDT and TEUGELS · Stochastic Processes for Insurance and Finance

ROSENBERGER and LACHIN · Randomization in Clinical Trials: Theory and Practice

ROSS · Introduction to Probability and Statistics for Engineers and Scientists

ROUSSEEUW and LEROY · Robust Regression and Outlier Detection

RUBIN · Multiple Imputation for Nonresponse in Surveys

RUBINSTEIN · Simulation and the Monte Carlo Method

RUBINSTEIN and MELAMED · Modern Simulation and Modeling

RYAN · Modern Regression Methods

RYAN · Statistical Methods for Quality Improvement, *Second Edition*

SALTELLI, CHAN and SCOTT (editors) · Sensitivity Analysis

*SCHEFFE · The Analysis of Variance

SCHIMEK · Smoothing and Regression: Approaches, Computation, and Application

SCHOTT · Matrix Analysis for Statistics

SCHOUTENS · Lévy Processes in Finance: Pricing Financial Derivatives

SCHUSS · Theory and Applications of Stochastic Differential Equations

SCOTT · Multivariate Density Estimation: Theory, Practice, and Visualization

*SEARLE · Linear Models

SEARLE · Linear Models for Unbalanced Data
SEARLE · Matrix Algebra Useful for Statistics
SEARLE, CASELLA and McCULLOCH · Variance Components
SEARLE and WILLETT · Matrix Algebra for Applied Economics
SEBER · Linear Regression Analysis
SEBER · Multivariate Observations
SEBER and WILD · Nonlinear Regression
SENNOTT · Stochastic Dynamic Programming and the Control of Queueing
Systems
*SERFLING · Approximation Theorems of Mathematical Statistics
SHAFER and VOVK · Probability and Finance: It's Only a Game!
SMALL and McLEISH · Hilbert Space Methods in Probability and Statistical
Inference
SRIVASTAVA · Methods of Multivariate Statistics
STAPLETON · Linear Statistical Models
STAUDTE and SHEATHER · Robust Estimation and Testing
STOYAN, KENDALL and MECKE · Stochastic Geometry and Its Applications,
Second Edition
STOYAN and STOYAN · Fractals, Random Shapes and Point Fields: Methods of
Geometrical Statistics
STYAN · The Collected Papers of T. W. Anderson: 1943–1985
SUTTON, ABRAMS, JONES, SHELDON and SONG · Methods for Meta-Analysis
in Medical Research
TANAKA · Time Series Analysis: Nonstationary and Noninvertible Distribution
Theory
THOMPSON · Empirical Model Building
THOMPSON · Sampling, *Second Edition*
THOMPSON · Simulation: A Modeler's Approach
THOMPSON and SEBER · Adaptive Sampling
THOMPSON, WILLIAMS and FINDLAY · Models for Investors in Real World
Markets
TIAO, BISGAARD, HILL, PEÑA and STIGLER (editors) · Box on Quality and
Discovery: with Design, Control, and Robustness
TIERNEY · LISP-STAT: An Object-Oriented Environment for Statistical
Computing and Dynamic Graphics
TSAY · Analysis of Financial Time Series
UPTON and FINGLETON · Spatial Data Analysis by Example, Volume II:
Categorical and Directional Data
VAN BELLE · Statistical Rules of Thumb
VIDAKOVIC · Statistical Modeling by Wavelets
WEISBERG · Applied Linear Regression, *Second Edition*
WELSH · Aspects of Statistical Inference

---

*Now available in a lower priced paperback edition in the Wiley Classics Library.

WESTFALL and YOUNG · Resampling-Based Multiple Testing: Examples and Methods for *p*-Value Adjustment

WHITTAKER · Graphical Models in Applied Multivariate Statistics

WINKER · Optimization Heuristics in Economics: Applications of Threshold Accepting

WONNACOTT and WONNACOTT · Econometrics, *Second Edition*

WOODING · Planning Pharmaceutical Clinical Trials: Basic Statistical Principles

WOOLSON and CLARKE · Statistical Methods for the Analysis of Biomedical Data, *Second Edition*

WU and HAMADA · Experiments: Planning, Analysis, and Parameter Design Optimization

YANG · The Construction Theory of Denumerable Markov Processes

*ZELLNER · An Introduction to Bayesian Inference in Econometrics

ZHOU, OBUCHOWSKI and McCLISH · Statistical Methods in Diagnostic Medicine